ARMS TRANSFERS TO ISRAEL

The Strategic Logic Behind
American Military Assistance

To my family and to the tribe

ARMS TRANSFERS TO ISRAEL

The Strategic Logic Behind American Military Assistance

DAVID RODMAN

sussex
ACADEMIC
PRESS
Brighton • Portland • Toronto

2 4 6 8 10 9 7 5 3 1

First published 2007, reprinted 2009, 2011, in Great Britain by
SUSSEX ACADEMIC PRESS
PO Box 139
Eastbourne BN24 9BP

and in the United States of America by
SUSSEX ACADEMIC PRESS
920 NE 58th Ave Suite 300
Portland, Oregon 97213–3786

and in Canada by
SUSSEX ACADEMIC PRESS (CANADA)
90 Arnold Avenue, Thornhill, Ontario L4J 1B5

British Library Cataloguing in Publication Data
A CIP catalogue record for this book is available from the British Library.

Library of Congress Cataloging-in-Publication Data
Arms transfers to Israel : the strategic logic behind American
military assistance / David Rodman.
p. cm.
Includes bibliographical references and index.
ISBN: 978-1-84519-178-8 (h/c : alk. paper)
1. United States—Military relations—Israel. 2. Israel—Military relations—United States. 3. Military assistance, American—Israel. 4. Arms transfers—Israel. 5. Arms transfers—United States. 6. United States—Military policy.
I. Title.

UA23.R61 2007
355′.0325956940973 22

2006032792

Typeset and designed by Sussex Academic Press, Brighton & Eastbourne.
Printed by TJ International, Padstow, Cornwall.
This book is printed on acid-free paper.

Contents

Illustrations

Illustrations, courtesy of the National Photo Collection, Government of Israel Press Office, are placed between pages 53 and 56.

Preface

The current strength of the American–Israeli arms relationship might tempt one to think that the Jewish state has always been a favored son of the United States in regard to weapons transfers. The historical facts, though, tell a different story. Despite repeated Israeli requests to purchase American arms, the United States consistently refused to furnish any major weapons systems to the Jewish state until the early 1960s. Not until the Kennedy administration concluded a deal to provide Israel with several Hawk anti-aircraft missile batteries in 1962 did the United States rescind its ban on arms sales to the Jewish state.

The real inauguration of an American–Israeli arms pipeline, however, occurred during the Johnson administration. Within the space of four years, this administration sold M-48 Patton tanks, A-4 Skyhawk aircraft, and F-4 Phantom aircraft to Israel. Indeed, by the time that the Johnson administration departed from office in the late 1960s, the United States had replaced Western Europe as the Jewish state's principal source of weapons.

Even though the American–Israeli relationship has often been marked by serious tensions since the time that this administration left office, the United States has faithfully remained Israel's main arms supplier. While the pursuit of incompatible national interests has led to frequent clashes between the United States and Israel, not only during Arab–Israeli wars, but also during periods when the Middle East has been characterized by relative calm, America has nevertheless delivered steadily greater quantities of increasingly sophisticated weapons systems to the Jewish state throughout the span of the past four decades.

The apparent contradiction between an expanding arms flow to the Jewish state, on the one hand, and recurrent problems between the United States and Israel, on the other hand, has given rise to two fundamental misconceptions about the arms component of the American–Israeli relationship. The first misconception maintains that arms transfers have been motivated primarily by American domestic politics rather than by American national interests. This line of thought views the power and influence of the American Jewish community – and the pro-Israeli lobby that serves as its spokesman – as the impetus behind weapons sales to the Jewish state. The second misconception, which is shared by many observers who endorse the first, insists that the United States has given a "blank check" to Israel – that the United States, in other words, has neither sought nor received any concessions from the Jewish state in exchange for weapons. Consequently, American national interests in the Middle East – and perhaps beyond – have never been served by these arms transfers.

The purpose of the present volume is to challenge these misconceptions about the arms component of the American–Israeli relationship. To this end, the book is divided into two parts. After the Introduction, whose underlying theme is the United States, Israel, and the Cold War, Part I begins in Chapter 1 by tracing the Israeli quest to acquire weapons in Western Europe and the United States. Chapters 2–4 chronicle the three large arms sales consummated with the Jewish state during the Johnson administration, as well as the temporary American arms embargo against Middle Eastern states in the aftermath of the 1967 Six-Day War. Chapter 5 then probes the cluster of reasons that drove the United States to provide weapons to Israel, demonstrating in the process that national interests, not domestic politics, were consistently uppermost in American decision making.

One caveat about Part I needs to be declared at the outset. The review of Johnson administration arms transfers does not delve deeply into the issue of bureaucratic infighting among the White House, State Department, and Defense Department. Chapters 2–4 sketch the strategic rationale behind arms sales to Israel; bureaucratic politics is introduced into the equation only insofar as it affected the debate over American national interests.

Part II addresses the myth that Israel has received American weapons free of political charge. Chapters 6–8 summarize and analyze the connection between American arms and Israeli concessions in the Six-Day War, the 1969–1970 War of Attrition, and the 1973 Yom Kippur War, respectively. Chapter 9 traces and assesses this connection in the absence of a full-scale Arab–Israeli war during the Nixon, Carter, and Reagan administrations. Collectively, these chapters reveal that a "security-for-autonomy" bargain has been at the heart of the American–Israeli relationship, a bargain in which the Jewish state has forfeited a certain amount of its autonomy in foreign policy decision making – more in wartime than in peacetime – in exchange for American arms. The claim that Israel has not had to pay a political price for American weapons, in short, is shown to be false.

Two caveats about Part II are in order here. The case studies in Chapters 6–9 employ the method of focused comparison; therefore, no effort is made to furnish comprehensive case studies of the

American–Israeli relationship in the Six-Day War, the War of Attrition, or the Yom Kippur War. Similarly, no effort is made to provide comprehensive case studies of the relationship during the Nixon, Carter, and Reagan years. Rather, the emphasis is strictly on the link between American arms transfers and Israeli diplomatic concessions in wartime and peacetime.

Second, and more importantly, because the case studies concentrate on the United States and Israel, one might be left with the impression that American and Israeli conduct were determined entirely by the behavior of the other party. That is, one might be left with the impression that neither American nor Israeli conduct was affected by Arab or Soviet behavior. The wartime and peacetime decision making of both the United States and Israel, in actuality, was clearly influenced by Arab and Soviet behavior. American decisions to supply or withhold arms at various points in the War of Attrition, to cite one example, were certainly affected by the extent of Egyptian and Soviet belligerence in the war. Likewise, the Israeli decision during the Yom Kippur War to refrain from crushing the Egyptian Third Army, to cite another example, was unquestionably influenced by the possibility of Soviet intervention, however remote, on behalf of Egypt. The fact that Arab and Soviet actions are not incorporated into the case studies, in sum, should not be taken to imply that they had no effect upon American and Israeli conduct. These actions are not included simply because they fall outside the scope of the case studies.

The Conclusion explains why both the United States and Israel have always preferred a patron–client relationship over an alliance. Simply put, each state has felt that the former constitutes a better arrangement than the latter with respect to the furtherance of its national interests.

Finally, it is necessary to say a few words about sources. This book relies mainly on American government documents from the *Foreign Relations of the United States* (*FRUS*) collection, from the Lyndon Baines Johnson Presidential Library, and from the United States National Archives. Few Israeli – and no Soviet or Arab – documents were consulted, because the book focuses on the American perspective on arms transfers to the Jewish state. An in-depth consideration of the Israeli, Soviet, and Arab perspectives is beyond its scope.

DAVID RODMAN
Dix Hills, New York

Acknowledgments

I would be remiss if I did not mention my indebtedness to a number of people who helped to turn this book into a reality. I would like to thank Efraim Karsh, in his capacity as editor of *Israel Affairs*, Barry Rubin, in his capacity as editor of *MERIA Journal*, and Sylvia Kedourie, in her capacity as editor of *Middle Eastern Studies*, for providing me with forums to publish some of the research that went into this book. Portions of the Introduction and Conclusion appeared in "Patron–Client Dynamics: Mapping the American–Israeli Relationship," *Israel Affairs*, Vol. 4, No. 2 (Winter 1997). Much of Chapter 2 and a portion of Chapter 5 appeared in "Armored Breakthrough: The 1965 American Sale of Tanks to Israel," *MERIA Journal*, Vol. 8, No. 2 (June 2004). Much of Chapter 4 and a portion of Chapter 5 appeared in "Phantom Fracas: The 1968 American Sale of F-4 Aircraft to Israel," *Middle Eastern Studies*, Vol. 40, No. 6 (November 2004). The journals division of Taylor & Francis Ltd. <www.tandf.co.uk/journals> has kindly permitted me to reuse material from my *Israel Affairs* and *Middle Eastern Studies* articles. The case studies in Chapters 6–8 build upon a chapter on the American–Israeli patron–client relationship in my earlier book, *Defense and Diplomacy in Israel's National Security Experience: Tactics, Partnerships, and Motives* (Brighton & Portland: Sussex Academic Press, 2005), and I would like to thank the press for allowing me to reuse some of the material there. The Government of Israel Press Office deserves thanks as well for permitting me to reproduce a number of the photos held in its National Photo Collection. Editorial Director Anthony Grahame and the staff at Sussex Academic Press deserve kudos for their help in turning a raw manuscript into a finished book. Finally, I would like to thank Harvey Cohen for carefully proofreading the galley pages. Any errors of fact or style that remain in the book, of course, are entirely my responsibility.

ARMS TRANSFERS TO ISRAEL

The Strategic Logic Behind
American Military Assistance

Introduction

The American–Israeli Relationship in Historical Perspective

Abba Eban, the late Israeli statesman who served as foreign minister during both the 1967 Six-Day War and the 1973 Yom Kippur War, once declared that:

> It is characteristic of American diplomacy that it seeks to accommodate a bewildering pluralism of objectives within every definition of its interests. It recoils from sharp, exclusive alignments and thus ends up by distributing displeasure across a broad field. Other nations complain that the United States is not a [one hundred] percent friend; but they also acknowledge that its adversarial postures are not intense and immutable. America's allies always have something to fear and its foes have something for which to hope.[1]

Eban, an astute observer of international diplomacy, had America's general approach to the Middle East in mind when he composed this comment; nevertheless, his insight applies equally well to the relationship between the United States and Israel.

Contrary to popular belief, the United States has not always been one hundred percent in Israel's camp since the establishment of the Jewish state in 1948. Rather, the American–Israeli relationship has deepened over the course of its existence, though not without numerous bumps along the way, with the 1960s witnessing perhaps the most significant evolution in its history. During this decade, the two states struck a "security-for-autonomy" bargain – a bargain that has been at the heart of the relationship ever since this time. Israel ceded a measure of its freedom over its own foreign policy to the United States in exchange for increased American backing. The United States enhanced its support of Israel in return for a measure of control over Israeli foreign policy.

Still, again contrary to popular belief, the American–Israeli relationship has never constituted an "alliance."[2] Rather, the relationship since the 1960s has been one between a patron, the United States, and a client,

Israel. The distinction between an alliance and a patron–client relationship is not merely semantic. An alliance is a formal agreement "between [or among] sovereign states for the putative purpose of coordinating their behavior in the event of specified contingencies of a military nature."[3] A patron–client relationship, on the other hand, does not rise to the level of a formal agreement. Neither the patron nor the client has any specific obligations or entitlements. The costs and benefits associated with an alliance, in short, are very different from those associated with a patron–client relationship.[4] The United States and Israel opted for the latter instead of the former based on their respective assessments of the costs and benefits attached to these alternatives. Each state, in other words, has felt that its national interests are better served by a patron–client relationship rather than by an alliance.[5]

This patron–client relationship did not emerge spontaneously upon the establishment of Israel. Throughout both the late 1940s and the entire 1950s, the American–Israeli relationship could not by any stretch of the imagination be labeled as one between a patron and a client. Indeed, a survey of these years reveals an often uneasy relationship, quite different from the one of later decades.

The Early Years of the American–Israeli Relationship

Throughout the years of the Truman administration (1948–1952), the American–Israeli relationship remained decidedly ambivalent.[6] The administration vacillated until the last moment on the issue of whether to accord recognition to Israel upon the Jewish state's birth in spring 1948. Against the advice of most of his advisors in the State and Defense Departments, President Harry Truman eventually granted *de facto* and, later, *de jure* recognition to Israel. Respect for the 1947 United Nations Partition Resolution, which mandated the creation of a Jewish state, the needs and desires of homeless Holocaust survivors, who wanted to rebuild their lives in a Jewish state, and the military victories of Israel in its 1947–1949 War of Independence, which showed that a Jewish state could survive in the midst of a hostile Arab world, finally swayed the administration's thinking on the issue of recognition. The Truman administration also sponsored Israel for membership in the United Nations, muted calls for the internationalization of Jerusalem, and provided some economic aid to the Jewish state.[7]

On the other hand, the administration pressed Israel to make significant concessions to the Arab world at the postwar Lausanne talks, particularly on the question of Arab refugee repatriation, though without any tangible results in the end. An American-backed arms embargo against the warring parties troubled Israel even more than the administration's diplomatic stance. Suffice to say that Prime Minister David Ben-Gurion's government felt that, by denying Israel the means to defend itself from Arab aggression, the United States was jeopardizing the Jewish state's very existence.[8] Moreover, the United States steadfastly refused to consider any sort of military commitment to Israel. Under no circumstances would it dispatch its own troops to fight on the new state's behalf.

At the outset of the Cold War, the United States had two fundamental, intertwined objectives in the Middle East: (1) to safeguard the West's oil supplies and (2) to prevent Soviet penetration of the area. Fear of Soviet penetration may even have played a role – but, if so, only a small one – in the Truman administration's decision to grant recognition and limited backing to Israel. The United States, after all, did not want the new state to turn to the Soviet Union for support: it did not want the Soviet Union to be able to use Israel as wedge to crack open the Middle East to "Communist" influence.

Principally, however, the administration viewed the Arab–Israeli conflict as an impediment in its quest to achieve its basic goals in the area. The conflict, therefore, had to be contained in a way that would not undermine American objectives in the region. While the United States would not abandon Israel, it would not embrace the Jewish state either. Instead, American foreign policy would generally incline more toward the Arab than the Israeli position on a final settlement of their conflict, because of the need to curry favor with an Arab world that controlled vital oil supplies and that had to be kept out of the Soviet sphere of influence.

The same essential logic guided the first Eisenhower administration (1953–1956) in its approach to the Middle East. Safeguarding the West's oil supplies and preventing the spread of Soviet influence in the area constituted this administration's main regional priorities. To this end, the United States took a leading part in putting together the 1955 Baghdad Pact. The administration believed that this alliance – which brought together Great Britain, Turkey, Pakistan, Iran, and Iraq – would form a solid barrier against Soviet penetration of the Middle East.[9] Ironically, though, the pact actually facilitated the spread of Soviet influence in the region. The Egyptian–Iraqi rivalry for dominance in the Arab world convinced Egypt to turn to the Soviet Union to counterbalance Western assistance to Iraq, particularly after the United States had spurned Egyptian requests for American arms. The Soviets promptly responded with a massive arms deal via Czechoslovakia.[10]

The first Eisenhower administration's attitude toward Israel reflected its emphasis on protecting oil supplies and keeping the Soviets at bay.[11] The United States sought to retain its influence in the Arab world by distancing itself from Israel. Hence, the American–Israeli relationship during these years was marked by frequent acrimony over such issues as border skirmishing and refugee resettlement. The administration prevented Israel from completing work on a water construction project along its northern border in 1953 by threatening economic sanctions. It also compelled the Jewish state to withdraw from the Sinai in the aftermath of the Israeli victory over Egyptian forces in the 1956 Sinai Campaign by threatening political and economic sanctions. The American–Israeli relationship became extremely tense during this last episode.[12] And, not surprisingly, the United States absolutely refused to provide either the arms or the security guarantee coveted by Israel.

But, as in the Truman years, this administration's conduct was not entirely opposed to Israeli interests. The United States continued to extend economic assistance to Israel.[13] It showed at least some understanding for the Israeli position with regard to an equitable sharing of water between the Jewish state and its Arab neighbors. It refused to endorse Great Britain's Alpha Plan, an initiative designed to strip Israel of much of the Negev in order to court the Arab world.[14] And, in exchange for an Israeli pullback from the Sinai, it committed itself in the wake of the Sinai Campaign to enforcing the right of Israeli shipping to free passage through the Straits of Tiran.

A genuine thaw in the American–Israeli relationship, however, would not occur until the years of the second Eisenhower administration (1957–1960).[15] American priorities in the Middle East had not changed suddenly. Rather, the Eisenhower administration revised its perception of Israel and the Arab world in light of developments in the region. First, the administration could not help but notice that Soviet influence in the Arab world had expanded dramatically in spite of America's chilly relationship with Israel. Egypt and Syria had moved into the Soviet orbit, and they would be joined by Iraq after the overthrow of its monarchical government in 1958. This expansion of Soviet influence had much more to do with rifts in the Arab world, particularly Egypt's struggle with conservative Arab states for leadership of that world, than with the Arab–Israeli conflict. Second, notwithstanding its condemnation of the Sinai Campaign, the administration came to see Israel as a robust military power on the basis of the elegant victory achieved by the Israel Defense Forces (IDF) in the war. Consequently, the United States believed that Israel could serve as a regional bulwark against further Soviet expansionism by helping to counter the efforts of radical, pro-Soviet Arab states like Egypt and Syria to destabilize conservative, pro-Western states like Jordan and Saudi Arabia.[16] Israel's assistance to the United States and Great Britain during the 1958 crises in Lebanon and Jordan, when it allowed American and British forces to pass through its airspace on their way to prop up the Lebanese and Jordanian governments, which were then being threatened by Egyptian-sponsored radicals, vigorously reinforced this view.[17]

The more positive attitude toward Israel, though, did not translate into public support for the Jewish state. The Eisenhower administration maintained the general ban on arms sales and refused to provide a security guarantee. Nor did it boost its economic or diplomatic backing of Israel. The changed attitude toward Israel manifested itself primarily in American restraint. In the late 1950s, and in contrast to its earlier conduct toward Israel, the Eisenhower administration no longer tried to coerce the Jewish state into making concessions to the Arab world, realizing that past efforts to do so had not gained it any credit in that quarter. To the Eisenhower administration's new way of thinking, then, the protection of the West's oil supplies and the containment of Soviet influence in the Middle East could not be bought with Israeli capital.

The Birth of the American–Israeli Patron–Client Relationship

Initially, the Kennedy administration (1961–1963) reverted to the pattern of behavior exhibited by the first Eisenhower administration with respect to Israel and the Arab world – that is, it tried to garner Arab, especially Egyptian, goodwill by holding out the prospect of major Israeli concessions to the Arab world (even though it never seriously contemplated the use of coercive tactics against the Jewish state).[18] Its motivation for acting so – to protect the West's oil supplies and to contain the spread of Soviet influence – reflected long-standing American goals in the Middle East. Solving the vexing problem of the Palestinian refugees appeared at the top of this administration's efforts to mend fences with Egypt.

The Arab-oriented thrust of American foreign policy, however, soon ground to a halt over Egypt's extreme behavior in the area. The Kennedy administration was very disturbed over Egypt's large-scale participation in the civil war then raging in Yemen, which included the use of chemical weapons against civilian targets. It also disapproved – albeit much less so – of the ongoing Egyptian campaign to destabilize Jordan.[19] Repeated American demands that Egypt moderate its conduct fell on deaf ears.

Once it realized that Egypt could not be dissuaded from its radical, pro-Soviet path, the administration concluded that moving closer to Israel would not do additional harm to American ties to the Arab world. Those Arab states that had pro-Western leanings, such as Jordan and Saudi Arabia, while they might resent a tighter American–Israeli relationship, would not turn to the Soviet Union in retaliation. Furthermore, radical Arab states, including Syria and Iraq, were already a lost cause.

The major upgrade in the American–Israeli relationship took the form of the Kennedy administration's repeal of the ban on heavy weapons sales to Israel. The United States eventually agreed to sell Israel several Hawk anti-aircraft missile batteries, which could be used to defend vital industrial and military installations, without any *quid pro quo* from the Jewish state regarding Palestinian refugees (or any other contentious issues related to the Arab–Israeli conflict).[20] Despite the fact that the Kennedy administration loudly proclaimed the sale to be a one-time exception to official policy, and not the inauguration of an arms pipeline to the Jewish state, an important taboo in the American–Israeli relationship had now been broken.

Although it is often suggested that the sale of Hawk missile batteries ushered in a new era in the American–Israeli relationship – that is, the era of the patron–client relationship – it is not really possible to cite any single event as the genesis of this relationship. Instead, the transformation in the relationship occurred more gradually throughout the 1960s. The Johnson administration (1963–1968), unlike its predecessors, never even attempted to engage in any sort of diplomatic flirtation with radical, pro-Soviet Arab states at Israel's expense.[21] More importantly, the one-time exception in arms sales to the Jewish state during the Kennedy administration disappeared during the Johnson administration. Not only did this administration enter into a number of major arms deals with Israel, but the United States had also become the Jewish state's principal source of weapons by the end of the decade.

Regardless of when precisely the American–Israeli patron–client relationship began in earnest, it is unquestionable that the transfer of American arms to Israel has been at its center from its inception. To understand why the United States moved from a position of refusing to supply arms to Israel to a position of providing large quantities of sophisticated weapons, as well as to understand how this military assistance has been manipulated to advance American national interests, it is first necessary to review the history of Israel's quest to arm itself in the face of unyielding Arab hostility to its existence.

PART I

How the Arms Relationship Emerged

1

The Israeli Quest for Arms

Western Europe and the United States

Moshe Dayan, the late Israeli soldier and statesman who served as defense minister in both the 1967 Six-Day War and the 1973 Yom Kippur War as well as foreign minister during Egyptian–Israeli peace talks in the late 1970s, once quipped that small states have no foreign policies; they have only, he said, defense policies. Perhaps Dayan's mindset is explained by the fact that Israel has fought no less than six full-scale wars to date and has been involved in low-intensity conflicts for much of the rest of its existence. One of the fundamental objectives of Israeli foreign policy has always been to ensure the state's security, and the procurement of arms from abroad to supplement local production has always been a basic priority in the search for security.

Arms Acquisition for a Jewish State

Indeed, the quest for arms long predates the establishment of the state. The Yishuv, the prestate Jewish community of Palestine, early on recognized that it would need weapons to defend itself against the surrounding Arabs, especially with the onset of anti-Jewish pogroms and violence in the 1920s.[1] The leadership of the community, therefore, decided to set up clandestine workshops to manufacture light arms and explosives, which would be stockpiled by the Haganah, the Yishuv's defense organization. Production of military-related goods for the British army during the Second World War years gave a significant boost to this underground industry. By the time of the 1947–1949 War of Independence, the Yishuv had a considerable arms-manufacturing capability with respect to light arms and explosives.

Nevertheless, the Jewish community of Palestine could not hope to manufacture all of the weapons that it knew would be necessary to survive a full-scale war with the Arab world. Heavy weapons – aircraft, armored vehicles, artillery tubes, and the like – would have to be acquired abroad, as would additional stocks of light arms, ammunition, and explosives. Consequently, the Yishuv actively sought to purchase weapons around the world, particularly in Europe and the United States.

The United States and Great Britain, working through the United Nations, however, slapped an arms (and manpower) embargo on the combatants during the War of Independence.[2] Intended to impede the war-making abilities of both Arabs and Israelis in order to encourage a quick end to the fighting on Western terms, the embargo should have favored the former, which already possessed armies equipped by their Western colonial patrons with the full range of heavy and light weapons (if not always in the greatest of numbers). Unintentionally, though, the embargo actually served to assist the Israeli war effort.[3]

American and British attempts to enforce the embargo around the world notwithstanding, both Arabs and Israelis sought to acquire weapons (and manpower) from abroad, especially in the United States and Europe.[4] Because Israel already had a sophisticated supply network in place around the world, it was far more adroit than the Arab world at getting around the embargo on arms (and men). Furthermore, rather than pursue arms (and men) overseas as aggressively as the Israelis, the Arab world seemed content to wait for its colonial patrons, particularly the British, to disregard the embargo, perhaps because these patrons had done so much for it in the past. Great Britain, for example, equipped and trained the Egyptian, Jordanian, and Iraqi armies before the war; it allowed Arab guerrillas to infiltrate into Palestine prior to the end of its mandate there, while preventing Jewish immigrants from entering the country; it handed over military-related goods and installations to Arab forces as it withdrew its army from Palestine; it permitted its officers to command Jordan's Arab Legion during the war; and it provided occasional fire support and intelligence data to Arab forces during the fighting.[5] By the time that the Arab world realized that its patrons were unwilling to violate openly the embargo on its behalf, it was simply too late to challenge seriously Israel's extensive efforts to circumvent the ban.

The weapons that allowed Israel to weather the initial Arab invasion in May 1948 and to fight the war to a victorious conclusion in January 1949 were supplied by the Soviet Union, albeit indirectly from the Czechoslovakian arsenal.[6] Crucial supplies of light arms and ammunition from this quarter permitted the Israel Defense Forces (IDF) to stifle the advance of Arab armies into the heartland of the Jewish state, while the heavy weapons that came afterward facilitated effective offensive operations later in the war.

From the 1950 Tripartite Declaration to the 1956 Sinai Campaign

To deter the outbreak of another round of full-scale Arab–Israeli warfare, the main Western powers in the region – the United States, Great Britain, and France – entered into the Tripartite Declaration. A key provision of this declaration concerned the prevention of an arms race between the Arabs and Israelis.

> The three Governments recognize that the Arab states and Israel all need to maintain a certain level of armed forces for the purposes of assuring their internal security and their legitimate self-defense and to permit them to play their part in the defense of the area as a whole. All applications for arms or war material for these countries will be considered in light of these principles. In this connection the three Governments wish to recall and reaffirm the terms of the statements made by their representatives on the [United Nations] Security Council . . . in which they declared their opposition to the development of an arms race between the Arab states and Israel.[7]

The Western powers, in other words, scrapped the arms embargo in favor of a mechanism for maintaining a balance of power through controlled weapons sales. This mechanism, they believed, would promote the Western interest of keeping the Soviet Union at bay in the Middle East, while not contributing to tensions that could result in war.

David Ben-Gurion, Israel's first prime minister, laid down the principle that the Jewish state should never be without at least one great power patron. A small state with minimal resources, he judiciously reasoned, Israel would need the support of a great power to ensure its security. After a short flirtation with the concept of nonalignment, the Jewish state linked its fortunes unambiguously to those of the Western world.[8] Once it did so, it concluded that the United States would be the ideal choice for a patron.

Most of all, Israel wanted access to American arms to correct what it perceived to be an unfavorable (to itself) balance of power in the Arab–Israeli arena. Throughout the early 1950s, Israel put forward a number of requests to purchase heavy weapons.[9] It sought to buy artillery tubes in 1953, aircraft in 1954, and aircraft, tanks, and artillery tubes in 1955 and 1956. Israeli requests became especially urgent after the announcement of the massive Czech–Egyptian arms deal in 1955, which threatened to undermine completely the existing local balance of power.

During the same years, but with much less enthusiasm, Israel also sought a security guarantee from the United States – that is, a formal American commitment to defend the Jewish state's territorial integrity, with its own troops if necessary. Israel strongly favored arms over a security guarantee because it feared that the United States would demand far-reaching Israeli concessions in return for the latter. Specifically, it was concerned that the United States would obligate it to cede too much control over its foreign policy, particularly with regard to its decisions to use force to defend its national interests, and would require it to acquiesce in arms sales to Arab states; therefore, Israel considered a security guarantee a second-best option.

Israel's fears about the concessions that it would have to make to get a security guarantee, however, were never realized, as the United States had no intention of making any kind of contractual defense commitment to the Jewish state. Nor did the United States intend to establish an arms pipeline to Israel. To justify its position, the first Eisenhower administration claimed that the balance of power already favored Israel in light of the superior quality of the IDF in comparison to the armies of the Arab states. It cited, in this context, the IDF's victories in repeated border skirmishes with the Egyptian, Jordanian, and Syrian armies. Moreover, the administration maintained that the United States did not want to get involved in bilateral military relationships with Middle Eastern states. Rather, it would only consider security guarantees and arms sales as part of some sort of regional arrangement. But the primary reason for the administration's reticence about a security guarantee or weapons for Israel, of course, concerned America's relationship with the Arab world. Simply put, the administration thought that providing Israel with a guarantee or selling it arms would harm the relationship between the United States and the Arab world, opening the way for Soviet expansionism in the Middle East and threatening the West's oil supplies in the area.

Consequently, Israel had to turn to Great Britain and France to meet its defense requirements.[10] In contrast to its effort to secure a security guarantee from the United States, Israel never seriously explored a similar arrangement with either Great Britain or France, believing that neither state constituted a reliable long-term patron. Indeed, when the British put out feelers about a defense treaty in 1951, the Ben-Gurion government rejected it.[11] Quite apart from the fact that the terms of the proposed treaty were weighted heavily in Great Britain's favor, Israel simply did not trust the British. The Jewish state wanted arms – and arms alone – from Great Britain and France in the early 1950s.

It managed to acquire some arms from Great Britain before the Sinai Campaign, including its first jet aircraft, some tanks, and two destroyers. Great Britain's willingness to sell weapons to Israel seemed to be predicated on a desire not to alienate the Jewish state entirely, as well as on a search for new markets for British arms. Its arms sales policy in the Middle East was far from favorable to Israel, though. The weapons transferred to the Jewish state were, at best, virtually obsolete, and, at worst, utterly dilapidated. And, in the two-to-three years before the Sinai Campaign, Great Britain largely stopped providing arms to Israel. Furthermore, Great Britain dispatched more arms to its traditional Arab clients, and sometimes these weapons were of better quality than those sent to the Jewish state.

Like Great Britain, France had interests in the Arab world, particularly in Syria and Lebanon, that initially made it wary of entering into a significant arms relationship with Israel. Thus, in the years prior to the Czech–Egyptian arms deal, it provided only modest quantities of aircraft, tanks, and artillery tubes to Israel. And, like Great Britain, it sold similar arms to various Arab states. Among its customers were Syria, Lebanon, and Egypt. By early 1956, however, France changed drastically its attitude about arms sales to Israel. It concluded an extensive deal in mid-1956 for aircraft, tanks, and artillery tubes – a sale that, while not quite matching the earlier Czech–Egyptian deal in numbers, went a long way toward redressing the weapons imbalance in the Arab–Israeli arena that had developed in the wake of Soviet arms supplies to the region. A further weapons deal in fall 1956 added substantially to Israel's arsenal of armored vehicles.

France's reason for furnishing arms to Israel on a large scale did not rest upon a newfound fondness for the Jewish state. Instead, Egypt's radical behavior in the Middle East, especially its fulsome support for rebel forces then beginning to challenge French rule in Algeria, convinced France to upgrade its arms relationship with Israel. France, in short, sought to build up the Jewish state as a counterweight to Egypt, thinking that, if it did so, the latter would not be able to meddle so freely in Algeria's internal problems.

While Great Britain vigorously objected to French arms sales to Israel, arguing that they undermined the purpose of the Tripartite Declaration, the Eisenhower administration tacitly accepted these deals, because they helped to offset Soviet weapons deliveries to Egypt. So long as Israel had an adequate supply of military equipment with which to protect itself, the United States had a ready made pretext to avoid making any defense commitment to the Jewish state.

From the Sinai Campaign to the Hawk Sale

Even though it had recently joined Great Britain and France in a war against Egypt, Israel still remained intent on acquiring American arms in the late 1950s. It also renewed its interest in an American security guarantee.[12] Yet, despite the fact that, in Lebanon and Jordan in 1958, Israel had assisted the United States to implement the Eisenhower Doctrine, under whose terms the latter had promised to lend support, including the dispatch of American troops, to states threatened by forces linked to "international Communism," the Jewish state did not achieve its objectives during the second Eisenhower administration.

Perhaps as a small reward for Israel's participation, the administration did authorize the sale of 100 recoilless rifles to Israel in 1958.[13] And it backed the Jewish state's efforts to obtain weapons in Great Britain and the Federal Republic of Germany. However, the one-time sale of a limited quantity of a light, defensive weapon by no means signaled an end to the general ban on arms sales to Israel, which stayed firmly in place. Moreover, the United States did not always support the Jewish state's weapons acquisition efforts in Western Europe, lest it become too closely associated with Israel. Despite a post-Sinai Campaign appreciation of the fact that a robust Israel could, in certain circumstances, contribute to the defense of American interests in the Middle East, the second Eisenhower administration continued to operate on the assumption that an overly close identification with the Jewish state would undermine its efforts to thwart Soviet expansionism and protect Western oil supplies in the region.

Unable to pry either arms or a security guarantee out of the United States, Israel again looked to Western Europe to meet its defense needs.[14] Their cooperation during the Sinai Campaign notwithstanding, the relationship between Great Britain and Israel did not experience a genuine relaxation of tensions until 1958, when the Jewish state permitted British troops to fly over its territory in order to bolster Jordan's pro-Western regime. In the aftermath of this event, and with American approval, the British began to furnish tanks to Israel, and they would continue to do so throughout the 1960s. Submarines were sold to the Jewish state as well. Great Britain's declining influence in the Arab Middle East, along with its growing belief that Israel could serve as a bulwark against Soviet expansionism in the area, propelled these sales.

France, though, constituted Israel's principal arms supplier from the late 1950s through the mid-1960s. Indeed, from the start of the Sinai Campaign until the end of the Algerian War in 1962, France and Israel entered into a patron–client relationship. The French furnished most of the Jewish state's aircraft and artillery tubes before the Six-Day War. Tanks, armored personnel carriers, anti-tank missiles, and air-to-air missiles were dispatched to the Jewish state, too. France and Israel also cooperated closely on the development of nuclear weapons and surface-to-surface missiles (SSMs).[15]

The Franco–Israeli patron–client relationship rested primarily on the mutual antipathy toward Egypt felt by the two states. The French were incensed by Egypt's backing of Algerian rebels who sought to expel France from their country, while Israel was concerned by Egypt's quest to unite the Arab world in an anti-Israeli crusade. Because the Franco–Israeli relationship lacked a positive basis, the Jewish state never felt comfortable with it. This discomfort would grow drastically as the Algerian War drew to a finish, when France no longer required Israel as a counterweight to Egypt. Once the war came to an end, in fact, the French moved quickly to upgrade their relationship with the Arab world. Though France would continue to provide arms to Israel before the Six-Day War, the patron–client relationship was essentially over by the early 1960s.

The fickleness of Great Britain and France as anything but temporary addresses for arms accounted in part for Israel's attempt to cultivate ties with the Federal Republic of Germany. Economic aid had long been flowing to the Jewish state as part of an attempt to atone for the Holocaust. Beginning in the late 1950s, with American approval, Israel also sought arms from the Federal Republic. From the late 1950s to the mid-1960s, the Federal Republic responded affirmatively to the Jewish state's requests for arms.[16] When this clandestine arms pipeline became public knowledge in 1965, however, the Federal Republic ended all weapons shipments to Israel, thinking that its diplomatic and economic position in the Arab world would collapse if it did not do so.

Concerns about the unreliability of Western Europe as a long-term source of arms meant that, once the Kennedy administration took office, Israel again turned to the United States for arms and a security guarantee.[17] While this administration neither entered into a steady arms relationship nor offered a binding security guarantee, it did put an end to the ban on sales of heavy weapons to Israel that had been in effect since the Jewish state's birth by permitting the purchase of several Hawk anti-aircraft missile batteries in 1962, opening the door to future deals.

Aside from the administration's calculation that the Hawk sale would not cause a further setback to America's relationship with the Arab world, it had a number of other reasons for going forward with the deal. First, in light of continued Soviet arms shipments to Egypt, Syria, and Iraq, the United States concluded that Israel had an authentic requirement for the Hawk batteries. Second, because the administration had put nuclear nonproliferation at the center of its global foreign policy agenda, and because Israel had already made significant progress toward acquiring nuclear weapons and SSMs, the United States thought that it could at least slow, if not actually prevent, the Jewish state from going nuclear by bolstering its security through the supply of a conventional weapons system.[18] And, third, a strengthened Israel would help to defend pro-Western Arab states by acting as a regional bulwark against the adventurism of pro-Soviet Arab states. Indeed, during the Jordanian crisis of the early 1960s, the Kennedy administration thought that Egypt's fear of an Israeli military response served as the most potent deterrent to open Egyptian intervention in the crisis.

But the major breakthrough in the American–Israeli arms relationship would wait until the Johnson administration took up the reins of office. In the space of less than four years, this administration entered into three major arms deals with Israel – it sold M-48 Patton tanks in 1965, A-4 Skyhawk aircraft in 1966, and F-4 Phantom aircraft in 1968.[19] How and why these sales took place is a complex story.

2

Armored Breakthrough

The 1965 Sale of M-48 Patton Tanks to Israel

In a 29 July 1965 exchange of letters between Deputy Assistant Secretary of Defense for International Security Affairs Peter Solbert and Special Assistant to the Defense Minister Zvi Dinstein, the United States and Israel reached a formal agreement on the sale of American tanks to the Israel Defense Forces (IDF). Under the terms of the agreement, the IDF would receive 210 M-48 Patton tanks. It would also receive conversion kits – to be used by it to replace the M-48's standard 90mm cannon with a more powerful and accurate 105mm cannon – as well as spare parts and ammunition.[1] The addition of these tanks to the Israeli arsenal would significantly boost the IDF's combat capabilities.

The sale of tanks represented a breakthrough in the American–Israeli relationship – and a very important one to boot. Unlike the Hawk missile, a strictly defensive weapon that could only be employed to protect air bases and other sensitive targets within Israel, the M-48 could be used in offensive warfare, to strike deep into Arab territory.[2] Not only did the M-48 constitute the first offensive weapon furnished to Israel, but its sale also set a precedent for the future. In contrast to the Hawk deal, which had not been quickly followed up with other arms sales, the United States agreed to supply Israel with 48 A-4 Skyhawk attack aircraft – another offensive weapon – not long after it had concluded the M-48 deal.[3] Of even greater consequence, the Johnson administration consented in November 1968 to furnish Israel with 50 F-4 Phantom aircraft.[4] A very powerful and sophisticated machine, the Phantom's capabilities far exceeded those of any warplane then in Arab arsenals. Collectively, these three arms sales signaled that the United States had chosen, albeit with some reluctance every step of the way, to become Israel's principal arms supplier, a considerable change from previous American arms policy.

The Johnson administration did not decide to sell tanks to Israel on the spur of the moment. Rather, it spent more than a year carefully considering the implications of such a deal for American national interests in the Middle East. Moreover, President Lyndon Johnson, Secretary of State Dean Rusk, and Secretary of Defense Robert McNamara – the administration's most senior officials in the realm of foreign policy – involved themselves closely in the debate surrounding the sale, as did the Joint Chiefs of Staff (JCS) and high-ranking Middle East experts in the National Security Council (NSC), the Department of State (DoS), the Department of Defense (DoD), and the Central Intelligence Agency (CIA). Clearly, the Johnson administration felt that a decision to supply Israel with tanks constituted a major initiative on its part.

The Road to the Sale

Israeli inquiries about tanks in the 1960s actually preceded the Johnson administration's rise to power. The idea of acquiring them had first been broached with the Kennedy administration in November 1963.[5] At the time, Israeli officials contended that the IDF would need 500 modern tanks, 300 during the next year and 200 more in two or three years, in order to offset the quantitative and qualitative advantage in armor possessed by the Arab "confrontation" states, mainly as a result of Soviet arms deliveries.[6]

The United States recognized the basic validity of this request, but it did not necessarily approve of the number of tanks desired by Israel. On at least three occasions, the JCS concluded that the IDF needed modern tanks to counter the expanding armored units of its Arab opponents.[7] The IDF had general military superiority over any combination of Arab armies, this body averred, but it risked falling behind in the preparedness of its armored units if it did not upgrade its tank inventory. So long as the IDF mothballed one obsolete tank for every modern vehicle it put into service – that is, so long as the IDF did not increase the total number of its tanks – American generals felt that its acquisition of new vehicles, specifically the M-48, would not unduly upset the local military balance in Israel's favor.

Civilian officials essentially concurred with the judgment of their military experts. By early 1964, Special Assistant to the President for National Security Affairs McGeorge Bundy had already acknowledged the legitimacy in principle of Israel's request for tanks.[8] McNamara echoed this sentiment by giving his blessing to a tank deal.[9] Even Rusk, never known to be a friend of Israel, seemed willing to offer tanks under the right set of circumstances.[10] Most importantly, Johnson believed that Israel required modern tanks. In a memorandum to his Deputy Special Council (for Jewish affairs) Myer Feldman, Johnson said, "On tanks specifically, we recognize Israel's armor needs gradual modernization to keep a dangerous imbalance from developing. . . . *We intend to see that Israel gets the tanks it needs. . . .*" [italics in original][11]

Nevertheless, during the first months of 1964, upon the advice of his advisors, Johnson demurred on a tank sale to Israel. With the prominent exception of Feldman, who argued for an immediate sale, none of them felt that the time had yet come to provide tanks.[12] Officials in the NSC, DoS, DoD, and CIA advanced several reasons to support their position. These reasons revolved around Israel's surface-to-surface missile (SSM) and nuclear research programs, expected Arab reaction to a tank sale, and traditional American arms policy in the Middle East.

Israel's development of a long-range SSM, in conjunction with its nuclear research agenda, caused much worry in the United States.[13] Military experts surmised (correctly) that Israel sought to acquire at least the capability to produce SSMs fitted with nuclear warheads in order to possess the "ultimate deterrent" to Arab aggression. A very expensive weapon to design and build, they knew, an SSM armed only with a conventional warhead made no sense from an economic or military point of view, even though American inspections of Israel's nuclear reactor at Dimona had uncovered no concrete evidence of a nuclear weapons program.

In their talks with Israeli counterparts, American officials repeatedly tied the prospects of a tank sale to Israel's "cooperation" on SSMs and nuclear weapons. American officials, quite simply, implied to their Israeli counterparts that a tank sale would be made contingent on an Israeli agreement to forgo SSMs and nuclear weapons. Johnson himself in a letter to Israeli Prime Minister Levi Eshkol made the connection, albeit in veiled terms.

> As you know, we have been giving careful thought to your expressed concerns about Israel's security needs. In particular we can understand your worries over the growing imbalance between Israeli and Arab armor, and can see the justification for your feeling that you must take steps to modernize Israel's tank forces and anti-tank defenses. We are fully prepared to discuss this problem further with you.
>
> At the same time we are disturbed lest other steps which Israel may contemplate taking may unnecessarily contribute to a heightened arms race in the region without in fact contributing to your security. Among other things, we seem to have quite different estimates with respect to the likely UAR [Egyptian] [surface-to-surface] missile threat, and the potential costs and risks of various ways of meeting it.[14]

Israeli leaders, not surprisingly, reacted quite angrily to this stance. They categorically rejected any connection between SSMs and nuclear research, on the one hand, and the acquisition of tanks, on the other hand. They had no intention of giving up the option to produce SSMs and nuclear weapons, even if it meant losing out on the opportunity to acquire American tanks, especially in light of the Johnson administration's refusal to translate its commitment to Israel's security into a formal guarantee.[15]

Another sticking point for the United States concerned Arab reaction to a tank sale. NSC, DoS, DoD, and CIA experts all believed that supplying Israel with tanks now would undoubtedly inflame Arab opinion against the United States and would quite possibly lead to serious setbacks for American national interests in the area.[16] These experts especially feared that Soviet influence among the Arabs would grow and that American oil concessions would be jeopardized.

Furthermore, American officials preferred to adhere for the time being to their traditional stance with regard to arms transfers to the Middle East – that is, the United States should avoid becoming a prominent supplier of weapons to either the Arab world or Israel. To this end, American officials still clung to the notion that they could reach some sort of understanding with the Soviet Union to restrict weapons deliveries to the Middle East in order to prevent an unrestrained Arab–Israeli arms race. American officials, in short, did not want to create a "polarized" Middle East, with the United States identified as Israel's benefactor and the Soviet Union identified as the Arabs' benefactor.[17]

The German Option to the Fore

The Johnson administration, therefore, decided to adopt a middle ground on a tank sale to Israel. While the United States would still refuse to sell tanks directly, it would employ its considerable influence in Western Europe to ensure that either Great Britain or the Federal Republic of Germany met Israel's requirements. Because the United States had long urged Israel to shop for weapons in Western Europe, the inclination to tell the Jewish state to look there did not constitute a radical departure from the past as concerned source of supply; however, never before had an American administration promised to exercise its diplomatic and financial clout so openly to fulfill Israeli arms needs.

The idea of an American-engineered sale of tanks to Israel from Western European states began to gather steam based on a joint DoS–DoD recommendation in the spring of 1964.[18] The Johnson administration quickly bought into the plan.[19] In his 1 June 1964 White House meeting with Eshkol, Johnson made it manifest that the United States would see that the IDF received tanks to offset its growing inferiority in armor. According to the minutes of the meeting:

> The President took up the specific problems on the agenda. With regard to tanks, he said he appreciated the readiness of Israel to agree to the manner in which tanks could be provided. He pointed out that we [the United States] could not provide tanks directly but we would be glad to help Israel in every way possible to get a sufficient quantity of tanks elsewhere.[20]

Significantly, the only *quid pro quo* that the United States demanded from Israel was secrecy. Though American officials would continue to express their concerns about Israel's SSM and nuclear programs – and would continue to pressure Israel on these programs in the future – they were no longer absolutely insisting on major Israeli concessions as the price of a tank deal.[21] The Johnson administration seemed to accept the notion that it could not compel Israel to give up its SSM and nuclear programs while Egypt refused to stop its own SSM and weapons of mass destruction programs.[22] Still, the United States could at least wield a restraining influence over the Eshkol government's decision making on these arms if it assisted in the acquisition of conventional weapons for the IDF.

The Johnson administration proposed two options to solve Israel's dilemma: purchase additional Centurion tanks from Great Britain or purchase M-48 tanks from the Federal Republic of Germany. The IDF already possessed the Centurion (and would buy more in later years), but it strongly preferred the M-48, which had greater "autonomy" than the Centurion – that is, it could operate on the battlefield for a longer period before having to rearm and refuel. The Eshkol government also preferred the M-48 for political reasons: the purchase of this tank would help to establish an American–Israeli arms pipeline, however roundabout, which Israel could later strengthen. Thus, Israeli officials pressed for the German alternative. They expressed discomfort, of course, that the Jewish state had to acquire American tanks indirectly, but they determined that the importance of getting these tanks took precedence over the actual source of supply.[23]

Despite the fact that the Federal Republic of Germany had been quietly sending arms to Israel since the late 1950s, it initially balked at providing M-48s.[24] German officials thought that furnishing a large number of tanks could not be kept secret for very long. Once the arms relationship became public knowledge, they feared, the Arab world would retaliate against West Germany. The Arab world might well seriously hinder, or even entirely sever, the lucrative economic ties that existed between itself and the Federal Republic. Of even more concern, the Arab world might decide to recognize formally the German Democratic Republic (East Germany), a development that the Federal Republic sought desperately to avoid.[25]

Nevertheless, after complex, three-way negotiations among American, Israeli, and German officials – negotiations during which the United States placed significant pressure on West Germany – the Federal Republic agreed to supply Israel with 150 M-48s. The United States agreed to restock West Germany's arsenal with a more modern variant of the M-48. It also agreed to furnish Israel with upgrade kits, spare parts, and ammunition, but not before some hard bargaining over prices and delivery schedules. All three states pledged to maintain strict secrecy about the existence of the deal.[26]

The Johnson administration facilitated this tank sale in part because its efforts to reach an accommodation with the Soviet Union on arms limitations in the Middle East appeared to be going nowhere. On a number of occasions, Rusk had instructed American diplomats to impress upon their Egyptian counterparts that the United States was under heavy pressure to supply Israel with arms in order to correct "imbalances" caused by large-scale Soviet deliveries of sophisticated weapons to Egypt (and, presumably, to other Arab states).[27] American diplomats relayed the message, but it apparently had no effect on either Egyptian or Soviet conduct.[28] Moreover, the Johnson administration had early indications that Jordan would soon be seeking more and better American arms than it had received in the past.[29] Given its commitment to Israel's security, the United States could not stand aloof as the Arab world acquired increasingly sophisticated weapons in ever greater quantities. In addition, the administration realized that a major Jordanian arms initiative would spell trouble domestically if Israel were not "compensated" in some fashion.

Predictably, the tripartite tank deal did not stay secret for very long. By October 1964, the American and West German media had got wind of the arrangement. How it leaked to the media has never been explained satisfactorily, though suspicions fall most strongly on disapproving officials within the West German government. Whatever the truth of the matter, Chancellor Ludwig Erhard announced in February 1965 that the Federal Republic of Germany intended to cease arms shipments to Israel immediately.[30] Only 40 tanks had reached the IDF by the time of the cutoff. The Eshkol government, naturally, turned to the Johnson administration, arguing that the United States should now supply tanks directly, as Israel's needs could no longer be satisfied by Western European sources.[31]

Water Woes and Arms Flows

While the tripartite tank deal unraveled under the glare of intense publicity, the Johnson administration confronted two more problems that drove it further down the road to direct weapons sales to Israel: the Arab world's Jordan River diversion project and Jordanian arms requests. The first problem was not a new one. Back in 1953, President Dwight Eisenhower had dispatched Ambassador Eric Johnston to the Middle East in order to reach an equitable agreement between Arabs and Israelis over the division of the Jordan River's water.[32] After several years of tortuous negotiations, Johnston worked out an informal understanding on how to apportion the water called the Unified Plan. But, because it would have implied recognition of Israel, the Arab world refused to sanction an official accord. Despite the lack of a formal agreement, Israel proceeded to build its National Water Carrier to implement its plan to use its share of the river's water to irrigate agricultural land.

There matters stood until 1964, when the National Water Carrier was set to begin pumping water throughout Israel. To abort this irrigation plan, the Arab world decided to divert the Jordan River's water from flowing into Israel, claiming (falsely) that the pumping would undermine its water rights. In a Knesset speech delivered on 21 January 1964, Eshkol warned in response that "Israel will oppose unilateral and illegal measures by the Arab states and will act to protect its vital interests."[33] Though force had not been threatened openly, the Israeli government had certainly signaled that it would resort to the military instrument if necessary. It reiterated this sentiment in a September 1964 communiqué.[34]

The United States strongly supported Israel's right to draw water in order to irrigate farmland. It steadfastly opposed, however, the use of force to protect this right, contending that Israel must take its case to the United Nations Security Council.[35] The Johnson administration worried that a flare up over the Arab diversion project had the potential to escalate into a full-scale Arab–Israeli war. This fear was heightened by Israeli–Syrian border skirmishing, which began in late 1964, over diversion-related engineering works on the Golan.[36] For its part, the Eshkol government regularly tried to reassure the Johnson administration that it would show restraint in the face of Arab provocations, but it consistently refused to forswear the use of force to defend its water rights under Johnston's Unified Plan.

The United States viewed the Jordanian arms issue as an even more urgent dilemma. Ostensibly, Jordan required additional arms, including tanks and aircraft, to offset the growing strength of the IDF. Actually, as Jordanian officials privately acknowledged and as American officials noted, it needed them primarily to resist pressure from Egypt to acquire Soviet arms.[37] Indeed, in less-than-subtle remarks, Jordanian officials told their American counterparts that Jordan would have to adopt Soviet arms if American weapons were not forthcoming.[38]

While the Johnson administration sought to keep Jordan firmly within the American sphere of influence, it also concluded that meeting King Hussein's demands in total would generate dangerous tensions in the Middle East. Not only would it undermine the administration's policy with respect to limiting weapons sales to the region, but it would also put the United States under tremendous pressure to compensate Israel by establishing a direct arms pipeline. An overt arms relationship with Israel, in turn, would push the Arab world more firmly into the Soviet orbit, creating a polarized Middle East and threatening American oil interests.[39]

Faced with the prospect of a very difficult choice, the Johnson administration decided on a middle-of-the-road approach. It would sell Jordan tanks, but not aircraft. Jordanian officials would be encouraged to shop in Western Europe for the latter. The administration felt that this compromise would serve its purposes on two fronts. First, it would keep Jordan within the American sphere of influence. Second, it could be "sold" to Israel, which would be told that its security was safeguarded better by an American-armed rather than a Soviet-armed Jordanian army, as well as that the United States had no choice but to supply arms to Jordan now that the American–Israeli–West German tank deal had become public knowledge.[40]

At first, Jordan did not react well to the American compromise proposal. Jordanian officials complained about the deal's scope: they wanted an advanced version of the M-48 tank as well as aircraft, whereas American officials had offered only the basic M-48. With regard to timing, they wanted the arms immediately, whereas American officials wanted to spread delivery out over a number of years. But, after some tough bargaining, Jordan essentially came around to the American point of view. In return for not seeking Soviet arms, the United States would supply 100 M-48 tanks. Furthermore, Jordan would defer for the time being its request to acquire more advanced M-48s and aircraft. Jordanian officials also promised that American tanks would not be deployed in Judea and Samaria.[41]

The Harriman–Komer Mission to Israel

The Johnson administration knew that a tank sale to Jordan would be met with protests from Israel's supporters within the United States if the Eshkol government objected to the deal. One influential official even argued that the United States needed Israel "not just grudgingly acquiescent *but actively in favor* [italics in original] of minimum arms aid to Jordan. Only if we can say [the Eshkol government is] 100% with us, can we protect our domestic flank."[42] Therefore, the Johnson administration had to reach some sort of accommodation with its Israeli counterpart. These negotiations proved to be far more complicated and contentious than the parallel American–Jordanian talks.

When first apprised of the prospective tank sale, Israeli officials reacted vehemently to the idea. The sale of tanks to Jordan, they asserted, would upset the Arab–Israeli balance of power by undermining the IDF's deterrent posture, particularly now that Israel had lost its West German source of arms. Furthermore, it would constitute a severe psychological blow to the Israeli people. Consequently, they went on, the IDF would be more inclined to take preemptive action in a future crisis.[43]

Still, the Eshkol government displayed a willingness to swallow a tank sale to Jordan – under the right circumstances. Israeli officials indicated that they would be willing to silence their opposition if (1) Jordan kept its American tanks on the East Bank of the Jordan River and (2) the United States agreed to sell arms to Israel directly.[44] The fact that some American officials had already resolved that it would be necessary to establish an arms pipeline to Israel in order to justify a tank sale to Jordan notwithstanding, the Johnson administration continued to evidence hesitancy in this regard.[45]

Into the tense atmosphere of the American–Israeli relationship in February 1965, Johnson dispatched Under Secretary of State for Political Affairs W. Averell Harriman and NSC staff member Robert Komer to the Middle East in order to hammer out a mutually acceptable agreement with the Eshkol government. Johnson, with the concurrence of Rusk, informed Harriman and Komer that they could tell the Eshkol government that the United States would consider "selective direct sales" of arms in the future, but only in exchange for a number of Israeli concessions. Not only must Israel quietly support American arms for Jordan by helping to mute opposition within the United States, but it must also give ironclad assurances that it would not develop nuclear weapons and would not take preemptive action against the Jordan River diversion project.[46]

Johnson and Rusk had to know that it was completely unrealistic to expect the Eshkol government to make these kinds of concessions in exchange for nothing more tangible than an American promise to consider direct arms sales. In any case, Israeli officials soon left the administration in no doubt as to their position in face-to-face talks with Harriman and Komer, stressing with great vigor that they could not enter into a one-sided agreement in which Israel made major concessions while the United States offered none of its own. As a counter to the American proposal, they said that they would acquiesce in an arms sale to Jordan if the United States supplied weapons directly to Israel. They also pledged that Israel would not undertake to manufacture nuclear weapons for the moment, though it would not give up the option to do so. And, on the Jordan River diversion project, they would only go so far as to say that Israel would exhaust peaceful means before resorting to the use of force.[47]

Harriman and Komer's recommendation that the Johnson administration obligate itself to direct arms sales, as well as soften its demands on nuclear weapons and the Jordan River diversion scheme, combined with the pressure to consummate an arms deal with Jordan before it turned to the Soviet Union, finally broke the impasse over an American–Israeli agreement.[48] Perhaps reluctantly, Johnson and Rusk acknowledged the logic of the Harrison–Komer position.[49] They opted for a limited agreement with the Eshkol government, largely on Israeli terms. On 10 March 1965, therefore, the United States and Israel entered into a Memorandum of Understanding (MoU). The central points of the agreement appear in clauses II and V:

> II. The Government of Israel has reaffirmed that Israel will not be the first [state] to introduce nuclear weapons into the Arab–Israel area.
>
> V. [The] United States will sell Israel on favorable credit terms, or otherwise help Israel procure, certain arms and military equipment as follows:
>
> > A. The United States will ensure the sale directly to Israel at her request of at least the same number and quality of tanks that it sells to Jordan.
> >
> > B. In the event of the Federal Government of Germany not supplying to Israel the remainder of the 150 M48 tanks outstanding under the German–Israeli tank deal of 1964, the United States will ensure the completion of this program.[50]

Other clauses of the MoU stipulated that Jordan would keep any tanks sold to it by the United States on the East Bank of the Jordan River and that the American–Israeli tank deal would remain secret until such time as the Johnson administration and the Eshkol government mutually agreed that it could be announced officially.

Before the American–Israeli tank deal could be completed and made public, the United States had to acquaint Arab states, especially Jordan and Egypt, with its existence.[51] The news, as expected, went down more easily in Jordan than in Egypt. The latter, after all, already had a strained relationship with the United States over its intervention in Yemen, not to mention its efforts to overthrow pro-Western Arab monarchies. Be that as it may, once the Johnson administration had formally informed Jordan and Egypt of the sale, and after a bit of last-minute haggling with the Eshkol government, the American–Israeli tank deal was officially concluded in the summer.[52]

3

One Step Forward and One Step Backward

The 1966 Sale of A-4 Skyhawk Aircraft to Israel and the Post-1967 Six-Day War Arms Embargo

On 22 February 1966 Deputy Special Assistant to the President for National Security Affairs Robert Komer informed President Lyndon Johnson that the United States and Israel had reached an agreement on the sale of American combat aircraft to the Israel Air Force (IAF).[1] Under the terms of the deal, the IAF would receive 48 A-4 Skyhawk attack aircraft. Though not the IAF's first choice among American aircraft, the addition of the Skyhawk to the Israeli arsenal meant that the air force's bombing capabilities would be improved greatly. Even if the aircraft were to arrive stripped of the advanced ordnance and sophisticated electronic systems then in the American inventory, it could still haul a large quantity of air-to-ground munitions over substantial distances.[2] The French combat aircraft that equipped the IAF's squadrons in the mid-1960s simply could not match its performance in this respect.

The significance of the Skyhawk sale extended beyond its contribution to Israeli airpower. Diplomatically speaking, the deal represented yet another step on the road toward the creation of a full-fledged American–Israeli patron–client relationship. The Skyhawk sale, consummated less than a year after the M-48 Patton tank deal, illustrated that the United States was well on the way to becoming Israel's principal source of arms.

As it did prior to the tank sale, the Johnson administration thought long and hard about a decision to furnish aircraft to Israel, carefully weighing the pros and cons of a deal. Almost one year passed between the date that it first agreed to consider an aircraft sale to Israel and the date of the actual deal. And, again, the administration's highest-ranking foreign policy officials – President Lyndon Johnson, Secretary of State Dean Rusk, and Secretary of Defense Robert McNamara – involved themselves intimately in the decision-making process, as did the Joint Chiefs of Staff (JCS) and experts from the National Security Council (NSC), Department of State (DoS), Department of Defense (DoD), and Central Intelligence Agency (CIA). As in the case of the tank deal, the Johnson administration felt that the sale of aircraft to Israel would have serious implications for American national interests in the Middle East.

The Israeli Wish List and the American Reaction

In its discussions with the Johnson administration over arms transfers, Israeli Prime Minister Levi Eshkol's government initially focused mainly on tanks, the item most urgently required by the Israel Defense Forces (IDF) in the mid-1960s. Still, aircraft had occasionally come up in talks between Israeli and American officials before the completion of the tank deal in the summer of 1965. Israeli officials had tentatively broached the issue of an aircraft sale as early as May 1964.[3] They raised the issue again – more forcefully this time – in February 1965 in order to signal to the Johnson administration that the IAF needed more and better aircraft to offset the increasing air strength of its Arab opponents.[4]

The Eshkol government, as a matter of record, had become insistent enough about aircraft that the 10 March 1965 Memorandum of Understanding (MoU) between the United States and Israel, which dealt primarily with the issue of an American tank sale, even contained a separate clause that referred specifically to aircraft. Clause V(C) said that: "The United States is further prepared to ensure an opportunity for Israel to purchase a certain number of combat aircraft, if not from Western sources, then from the United States."[5] This vague formulation satisfied both the Johnson administration, which had thus far shown no real inclination to sell aircraft to Israel, and the Eshkol government, which felt that the United States had finally acknowledged the IAF's requirement for new aircraft.

Israel displayed special interest in the F-4 Phantom, then a frontline aircraft with the United States Navy, Air Force, and Marine Corps. Designed to be a versatile aircraft, the rugged, fast, and modern Phantom could engage in air-to-air combat against interceptors as well as execute air-to-ground attack missions against all types of targets. Because the IAF had fewer aircraft at its disposal than its Arab opponents, it had traditionally sought flexible machines that could compensate for this numerical gap by undertaking different kinds of missions. Thus, the Phantom seemed a perfect fit for the IAF, despite its exorbitant price tag. That it could also deliver a very large tonnage of ordnance over a very long range made it that much more attractive to the IAF.

The Eshkol government, however, realized that Israel had no genuine hope of acquiring the Phantom in 1965. In a late February conversation with American officials, the prime minister sullenly accommodated himself to this reality.[6] Once the Phantom proved to be a nonstarter on diplomatic and economic grounds, Israel turned its sights on a dedicated strike aircraft that could carry a large bomb load to a distant target. The IAF wanted the A-6 Intruder, a frontline United States Navy aircraft that possessed an advanced electronic warfare capability to aid in its air-to-ground attack missions. The Skyhawk – which had a shorter range, a smaller ordnance capacity, and a less sophisticated electronic warfare package than the Intruder – was third on Israel's wish list.

For its part, the United States did not believe that the IAF had an immediate need for new aircraft in 1965. The JCS argued that the current Arab–Israeli air balance had not swung markedly in favor of the Arab world; hence, this body recommended against an aircraft sale to Israel at the time.[7] Even the sale of 24 aircraft, the number that the United States had bandied about in connection with the March 1965 MoU, it thought, would tip the air balance in favor of the IAF. The JCS also noted that, if the Johnson administration nevertheless found it absolutely necessary to sell aircraft to Israel, the IAF should not be supplied with attack aircraft. Instead, the United States, this body averred, should offer to furnish the F-5 Freedom Fighter, an interceptor that Israel had never requested (because the IAF had a sufficient number of French Mirage IIIC fighters).

The Road to the Sale

With the tank sale "under its belt," Komer informed Johnson in October 1965, the Eshkol government had now begun to concentrate on aircraft single-mindedly.[8] Administration officials argued that the terms of the March 1965 MoU did not obligate the United States to furnish aircraft to Israel – at least until the Eshkol government had made an exhaustive effort to purchase them in Western Europe. Only if this attempt came to naught would the United States then seriously consider an aircraft sale. Throughout the spring, summer, and autumn of 1965, therefore, American officials urged their Israeli counterparts to search Europe for aircraft.[9]

The Johnson administration believed that the IAF's demand for additional aircraft could be met by a purchase of the British Canberra or Buccaneer, or the French Vautour, the latter perhaps upgraded with a more powerful British-built engine.[10] In response, the Eshkol government claimed repeatedly that Israel had looked into acquiring suitable aircraft in Western Europe, but had met with no success in either Great Britain or France.[11] The IAF contended that neither the Canberra nor the Buccaneer measured up to Israel's performance needs for an attack aircraft. With respect to the Vautour, Israel claimed both that the production line for this aircraft had been shut down and that France could not part with any of its own aircraft.[12]

The United States responded skeptically to the Israeli position at first. IAF commander General Ezer Weizman's talks with a joint DoS–DoD delegation on 12–13 October 1965 did nothing to help the situation, especially after he made a "blue sky" request for 210 aircraft.[13] The Johnson administration and the Eshkol government, so it appeared, had reached an impasse on the issue of an aircraft sale.

But, in autumn 1965, the United States had already begun to rethink the idea of supplying Israel with aircraft, particularly after canvassing Western Europe on its own.[14] Komer hinted that a small sale during 1966 might be possible.[15] By January 1966, he implied that an aircraft deal had now become inevitable, only its timing and terms remained uncertain.[16] Not long afterward, Rusk and McNamara proposed a simultaneous sale of aircraft to Israel and Jordan, with the former receiving 48 Skyhawks and the latter receiving either 36 F-5s, 36 F-104 Starfighters, or a 36-plane package combining the two types.[17] Though the secretaries felt that the best alternative for the United States would be to reach an arms limitation agreement with the Soviet Union on weapons transfers to the Middle East, the next best alternative would be to make controlled arms sales to Israel and friendly Arab states.

Events moved at a swift pace from here. Komer laid out the logic of a simultaneous aircraft sale to Israel and Jordan in a long communication with Johnson.[18] In a subsequent meeting with Israeli officials, Johnson intimated, albeit in veiled language, that the IAF would get American aircraft.[19] The concrete terms of a deal were hammered out rapidly thereafter. Israel would not be permitted to obtain the Intruder, but rather would receive the Skyhawk.[20] The Eshkol government made one last gambit to get the Intruder, but it quickly settled for the Skyhawk.[21]

In return for the 48 Skyhawks, Israel would: (1) reaffirm its undertaking not to be the first state to introduce nuclear weapons into the Middle East; (2) reaffirm its undertaking to allow periodic American inspections of its Dimona nuclear research facility; (3) search Western Europe, not the United States, for the bulk of its arms in the future; (4) refrain from opposing the sale of aircraft to Jordan; and (5) maintain secrecy about the sale until the United States chose to publicize it.[22] Before the end of February, the Johnson administration and the Eshkol government had signed off on the deal.[23]

The Post-Six-Day War Arms Freeze

The burgeoning American–Israeli arms relationship, however, would be briefly interrupted just a few months later. When the Six-Day War erupted in early June 1967, the United States quickly slapped an official arms freeze on Israel and pro-Western Arab states, including Jordan, Saudi Arabia, Yemen, Morocco, Tunisia, and Lebanon. Though some consignments of military equipment already in the pipeline before the outbreak of war eventually continued on to their destinations, the Johnson administration refused to be drawn into negotiations for new weapons sales, particularly with respect to major items like tanks and aircraft. It would not even guarantee that all of the arms sold previously to its Middle Eastern customers would ultimately be delivered to their respective governments. For the Jewish state, the embargo placed in question the earlier Skyhawk sale.

The Johnson administration never enthusiastically embraced the role of weapons merchant. Thus, when it sensed that the Six-Day War might offer an opportunity to stem the flow of arms to the Middle East, it sought to take advantage of the situation. In the years prior to the war, the United States had expressed a desire to enter into some sort of regional arms limitation agreement with the Soviet Union. The latter, though, had shown no genuine interest at the time. The crushing battlefield defeat suffered by the Soviet Union's main regional clients – Egypt, Syria, and Iraq – during the war, American officials hoped, would lead to a change of heart among their Soviet counterparts.

Indeed, while the fighting still raged on, the Johnson administration had already started to consider the prospect of negotiating a postwar Middle Eastern arms limitation agreement with the Soviet Union.[24] Once the guns fell silent, it intensified its efforts, suggesting that an accord also incorporate a United Nations (UN) arms registry so that all future weapons transfers to the region would become public knowledge.[25] On 19 June 1967, Johnson himself made Middle Eastern arms control, including a UN registry, one of his five principles for a comprehensive Arab–Israeli settlement in a speech that received widespread exposure at home and abroad.

The United States initially pressed ahead with its arms limitation agenda, despite the fact that three of the "big four" powers did not endorse its position. Neither the French nor the British, American officials conceded soon after the war, seemed overly enthusiastic about a Middle Eastern arms control agreement.[26] For their part, the Soviets replied with barely concealed scorn to the American proposal. Throughout June and July 1967, repeated attempts by American officials to interest their Soviet counterparts in restricting weapons deliveries to the region met with a frosty refusal to engage in a serious discussion of the issue.[27] By the end of the summer, therefore, the Johnson administration realized that a Middle Eastern arms control agreement simply was not on the horizon.[28]

The United States had not only to contend with West European disinterest and Soviet opposition, but also with Israeli and Arab demands for an end to the arms freeze. Just a few days after the end of the Six-Day War, Israeli diplomats began to voice their concern about Soviet rearmament of Arab states, even though they generally concurred with the American assessment that this resupply effort was intended to reassert influence in the region rather than to prepare the Arab world for another round of warfare.[29] By mid-July, the Eshkol government had become alarmed enough about the Soviet arms transfers to ask American officials to allow an Israeli military mission to travel to the United States to discuss the regional balance of power and the IDF's future weapons requirements.[30] For the rest of the summer and into the fall, Israeli officials would become ever more insistent that the Johnson administration lift the arms ban, suspecting that the freeze was being deliberately manipulated in order to influence the Eshkol government's postwar bargaining position with regard to a settlement of the Arab–Israeli conflict.[31]

On the eve of the Six-Day War, the Eshkol government had sought arms from the United States in the form of 140 M-60 tanks (as a substitute for 140 M-48s that were then undergoing overhauls in Israel and, hence, were unavailable to the IDF), one additional Hawk missile battery plus 100 more missile rounds, and expedited delivery of 24 of the previously promised 48 Skyhawks, the first of which were due for delivery in December 1967.[32] Israel, however, had to settle for a limited quantity of gas masks, which the Johnson administration consented to supply in light of Egypt's known possession and prior employment of chemical weapons.[33]

In the wake of the IDF's spectacular victory in the Six-Day War, Israel had reassessed its immediate arms requirements. The destruction of the Egyptian, Jordanian, and Syrian armies, coupled with the large numbers of tanks, armored personnel carriers, and artillery tubes captured by the IDF, meant that Israel was no longer worried about land force ratios in the short run. It would take years for the Arabs to rebuild their shattered armies to the point where they could once again pose a real challenge to the IDF.

The Jewish state, on the other hand, was far less sanguine about the air balance. The Soviet Union was dispatching replacement aircraft to Egypt, Syria, and Iraq at a steady pace in order to make up for their wartime losses. The IAF, which had lost about 20 percent of its inventory in the war, to the contrary, had not received any new aircraft since the outbreak of fighting. Though France – Israel's sole source of planes from the mid-1950s to the mid-1960s – continued to deliver spare parts for the IAF's existing fleet of aircraft, it had shown absolutely no inclination to sell replacement aircraft. An Israeli military mission made clear to American officials that the IAF urgently required new aircraft when it requested that the United States provide on schedule the previously promised 48 Skyhawks as well as supply 27 additional Skyhawks and 50 Phantoms by the end of 1968.[34]

Jordan, too, began to press the United States to lift the arms freeze not long after the Six-Day War. Jordanian officials hinted that they might have no alternative but to turn to the Soviet Union for weapons if the United States did not prove forthcoming.[35] Like their Israeli counterparts, they would continue to present arms requests to American officials throughout the summer and into the fall.[36]

The Demise of the Arms Freeze

Less than a month after declaring the weapons ban, American officials were already beginning to entertain second thoughts about its utility. By early summer, faced with a flat Soviet refusal to reciprocate American restraint, as well as rapidly rising Israeli and Arab anger, DoS and DoD officials believed that it would be a good idea to furnish Israel and pro-Western Arab states with the arms that the United States had consented to deliver before the Six-Day War.[37] By early August, the Johnson administration approved the sale of some ammunition and spare parts to Israel, as well as a few "nonlethal" items to pro-Western Arab states.[38]

Nevertheless, the Johnson administration was not yet ready to promise either Israel or friendly Arab states that the United States would stand behind existing contracts for big ticket items like aircraft. Nor would the administration agree to negotiate new arms deals. It would not be until late October, as a matter of fact, that the United States consented to honor all outstanding contracts.[39] Israel would get the Skyhawks on schedule and a number of pro-Western Arab states, excluding Jordan for the moment, would receive "compensatory" arms.

The Johnson administration appears to have delayed until the last days of October the announcement that the Skyhawk deal would go forward for two principal reasons. First, the United States felt that Israel had come through the Six-Day War in strong shape. Both American intelligence officials and the JCS argued that, despite its losses in the war, Israel's strength *vis-à-vis* its Arab opponents had grown considerably as a result of the fighting.[40] Not only had Arab losses been much heavier, but the Arab world now also recognized its military inferiority in relation to the Jewish state. The United States did not think that the Arabs would have a viable war option for at least 12–18 months, the Soviet resupply effort notwithstanding. Israel, in short, could defend itself adequately for the time being with the arms currently at its disposal.

Second, the Johnson administration wanted the Eshkol government to help convince Congress not to impede American foreign aid programs now that the Soviet Union had rejected the idea of Middle Eastern arms control. The administration feared that, if Congress demonstrated a reluctance to authorize arms sales to pro-Western Arab states, especially Jordan, they would then turn to the Soviet Union to fill their weapons requirements. During the summer and fall, therefore, American officials urged Israeli diplomats to use their influence on Capitol Hill, via the American Jewish community, to lobby on behalf of the administration's intention to conclude weapons deals with friendly Arab states.[41] The arms freeze, of course, gave American officials tremendous clout on this issue with their Israeli counterparts.

For its part, the Eshkol government agreed to do the administration's bidding on Capitol Hill, even though it certainly was not keen on an American arms pipeline to Arab states, particularly Jordan.[42] The Eshkol government not only greatly desired the repeal of the arms ban, but it also grudgingly accepted the Johnson administration's logic that Israel itself would be better off if Arab states, including Jordan, were to be equipped with American weapons rather than Soviet arms. The Jewish state, after all, much preferred American to Soviet influence in the Arab world.

While the Johnson administration used the arms freeze to manipulate Israeli behavior on the issue of weapons sales to friendly Arab states, the Eshkol government's anxiety that the ban would be employed to elicit other concessions turned out to be overblown. The United States released the Skyhawks without further demands.

For all intents and purposes, the American weapons ban came to an end in late October, with the Johnson administration's public announcement that Israel and pro-Western Arab states would receive the arms that they had already been promised by the United States. If conclusive evidence were needed of the freeze's death, however, then the delivery of the first four Skyhawks to Israel in December served as proof. The Johnson administration's experiment with arms control had collapsed under the weight of Middle Eastern realities. The ban, which would have harmed Israel more than the Arab world had it been lengthened beyond its short life span, did not result in any tangible consequences for either Israeli or Arab security.

4
Air Support
The 1968 Sale of F-4 Phantom Aircraft to Israel

In a late November 1968 exchange of letters between Assistant Secretary of Defense for International Security Affairs Paul Warnke and Israeli Ambassador to the United States Yitzhak Rabin, the Johnson administration formally agreed to supply Israel with 50 F-4 Phantom aircraft.[1] Under the terms of the sale, the Israel Air Force (IAF) would also receive associated ordnance, spare parts, maintenance equipment, and training. Six of the Phantoms would be specially configured to serve as reconnaissance platforms, though they would be capable of fulfilling standard combat missions as well. The first batch of aircraft would reach Israel in September 1969, according to an accelerated delivery schedule – one that would not, however, interfere with Phantom production for American military forces.[2]

On both military and diplomatic grounds, the Phantom sale held great significance for Israel. Militarily speaking, the addition of this state-of-the-art aircraft to the Israeli order of battle massively upgraded the IAF's fighting power. Diplomatically speaking, the deal signaled an enhanced American commitment to Israel's security. Likewise, the sale had very important implications for American foreign policy. It meant that the United States had now chosen, in effect, to become Israel's principal arms supplier, a role that it had strenuously sought to avoid before the 1967 Six-Day War. Furthermore, at least insofar as concerned airpower, the United States had now also reconciled itself to providing Israel with better arms than those in the hands of its Arab opponents, another departure from past policy.

Not every Johnson administration official necessarily viewed the Phantom sale as quite so consequential at the time, but they all nevertheless recognized that it could seriously affect America's standing in the Middle East by infuriating the Soviet Union and the Arab world. For this reason, the administration thought long and hard about its decision to sell the Phantom to Israel, meticulously weighing the potential benefits and potential costs of such a deal. Indeed, it took almost all of 1968 to make up its mind on the sale. Moreover, the administration's most senior foreign policy officials – President Lyndon Johnson, Secretary of State Dean Rusk, and Secretaries of Defense Robert McNamara and Clark Clifford – regularly inserted themselves into the decision-making process. High-ranking officials from the Joint Chiefs of Staff (JCS), Department of State (DoS), and Department of Defense (DoD) expended a considerable amount of energy on the Phantom deal, too.

The Arab–Israeli Air Balance

The Phantom had been at the top of Israel's shopping list long before the Six-Day War, but the Johnson administration had no intention of selling this aircraft to the Jewish state in those years. Israel simply did not need the Phantom, the administration argued, because the IAF already possessed the means to hold its own in battle. The addition of this aircraft to the IAF's arsenal, administration officials continued, would give Israel a tremendous military advantage over the Arab world – an advantage, they concluded, that would not be helpful to furthering American national interests in the Middle East. There the matter of the Phantom rested until the end of the Six-Day War.

Though the IAF had achieved an astonishing victory in that war – one that had witnessed the virtual annihilation of the Egyptian, Syrian, and Jordanian air forces – the Eshkol government soon voiced grave concerns about the postwar air balance of power.[3] True, the Arab air forces, Israel acknowledged, had not yet recovered entirely from their collective defeat; however, in the first six months following the war, the Soviets had already poured large numbers of replacement aircraft into the Egyptian, Syrian, and Iraqi arsenals. The Soviets had also begun to help Egypt, Syria, and Iraq to construct more and better airfields, which would make their air forces less vulnerable to the IAF in the future. The Soviets had stepped up their training efforts on behalf of Arab air and ground crews as well. Moreover, Soviet combat squadrons had been spotted in Egypt.

Meanwhile, the Eshkol government contended, the IAF's strength had been seriously sapped by the Six-Day War. The air force had lost 20 percent of its frontline aircraft in the fighting, and many of its older jets were rapidly approaching obsolescence.[4] Furthermore, Israel no longer had the prospect of acquiring replacement aircraft in Western Europe, its primary source of supply before the war. Arab air forces already outnumbered the IAF by a ratio of at least three to one – and this ratio would only get worse as the months wore on. Even the delivery of the 48 Skyhawks purchased back in early 1966, which were scheduled to enter IAF service in meaningful numbers during 1968, would not suffice to solve the problem. The IAF, the Eshkol government asserted, would require 250 modern aircraft in the late 1960s, with the total increasing to 350 during the early 1970s, in order to offset the Arab air threat.[5] The IAF, in short, needed additional modern aircraft immediately, and the United States constituted the sole arms supplier available to Israel.

Predicated on this assessment, the Eshkol government approached the Johnson administration with three requests.[6] First, it asked the United States to expedite the delivery of the 48 Skyhawks already on order for the IAF. Second, the Eshkol government asked to purchase another 27 Skyhawks, to bring the total to 75 aircraft. Third, and most important from the Israeli perspective, it asked to purchase 50 Phantoms, thereby officially renewing its interest in this aircraft.

To make its case for these requests, Israel naturally painted a worst-case scenario of the IAF's predicament. For its part, the United States partially endorsed the Israeli assessment of the air balance.[7] It essentially agreed with Israeli estimates of the relative numbers of aircraft that would be in the hands of the IAF and the Arab air forces over time. It also saw merit in the linked claims that a considerable portion of the IAF's inventory was quickly nearing obsolescence and that the IAF faced a crippling shortage of modern aircraft in the future.

Be that as it may, buoyed by the expert opinion of the JCS, the Johnson administration had a decidedly more optimistic overall assessment of the air balance.[8] General Earle Wheeler, chairman of the JCS, spoke for the administration when he argued that, at a minimum, the IAF would remain superior to the combined Arab air forces for the next 18 months – that is, until the middle of 1969. American officials, in fact, argued that the IAF had actually grown stronger in relation to these air forces since the Six-Day War, particularly now that the Skyhawk had started to join the Israeli fleet.

The administration set forth several reasons to support its position. First, Israel had completely ignored such intangibles as intelligence, command and control, maintenance and logistics, personnel skills, training, and morale – areas in which the IAF had tremendous advantages that could not be easily overcome by the Arab air forces. Second, though plainly worried by the increased Soviet military presence and activity in the Middle East, American officials were not as alarmed about these developments as their Israeli counterparts. The administration did not feel that the Soviet Union would become directly involved in Arab–Israeli hostilities. Nor did it necessarily agree with the Eshkol government that the Soviet Union intended to build up Arab arsenals to levels far beyond their pre-Six-Day War limits. And, third, in contrast to their Israeli counterparts,

American officials thought that the IAF might yet be able to obtain aircraft in Western Europe.[9]

The Road to the Sale

On the basis of this assessment, the Johnson administration determined that it would not provide the Phantom to Israel when Eshkol visited the United States for talks in January 1968. Rather, it would be prepared to discuss the sale of additional Skyhawks as insurance against a short-term deterioration in the Arab–Israeli air balance. A decision on the Phantom would be held in abeyance until several trends in the Middle East, including Soviet and French arms transfer policies as well as progress in Arab–Israeli peace negotiations, came into clearer focus over the course of 1968. The administration's key decision makers – Johnson, Rusk, McNamara, Under Secretary of State Nicholas Katzenbach, and Special Assistant to the President Walt Rostow among them – all approved of this plan.[10] Ambassador to Israel Walworth Barbour represented the lone dissenting voice, asserting that it would be best for the United States to meet Israel's aircraft requirements as promptly as possible.[11]

On 7–8 January 1968, Johnson, Eshkol, and their aides held three meetings. The topic of additional aircraft for Israel occupied a prominent place in all of them.[12] The Eshkol government and the Johnson administration made their respective cases for and against an immediate Phantom sale. During the last meeting, Johnson expressed a willingness to furnish Israel with up to 40 additional Skyhawks; but, a decision on the Phantom, he said, would depend on developments in the Middle East over the coming year. He did promise, however, that Israel would receive its first Phantoms no later than January 1970 if the United States eventually chose to supply the aircraft to the IAF – any delay in making a positive decision on the Phantom, in other words, would not delay its arrival in Israel.[13]

On 30 January 1968, the United States formally agreed to sell 40 additional Skyhawks to the Jewish state.[14] On 27 September 1968, following the advice of Rusk, Clifford, and Rostow, Johnson consented to the sale of an additional 12 Skyhawks, to bring the total provided to Israel to 100 aircraft, the last of which would be delivered to the IAF by early 1970.[15] These deals constituted concrete proof of America's ironclad commitment to Israel's security in the eyes of the Johnson administration, which hoped that they would alleviate some of the urgency over a Phantom sale.

Israeli officials had absolutely no intention of reducing their pressure on the administration to sell the Phantom, though. Less than two weeks after the Johnson–Eshkol talks, they were already asking their American counterparts for clarifications on where the United States stood on the sale of this aircraft.[16] By the end of February, Israel again emphasized its need for more and better American aircraft, calling attention once more to the increased Soviet presence in Egypt.[17] In late April, Israeli officials made another plea for the Phantom.[18] A few days later, Rabin conveyed a personal letter from Eshkol to Johnson asking for an immediate positive response on the sale of the Phantom.[19] Israeli officials kept up their efforts to convince the administration to sell the Phantom throughout the summer and fall.[20] Indeed, the Eshkol government maintained steady pressure on the United States until the moment when the Johnson administration indicated its readiness to begin negotiations on the sale of the aircraft to Israel.

Throughout the first half of the year, the Johnson administration parried the Eshkol government's repeated requests for the Phantom by asserting that the United States could not consider a sale until Soviet and French arms policies in the Middle East had become perfectly clear. Despite the fact that the Soviet Union had not responded in kind to the administration's freeze on weapons transfers to the Middle East in the wake of the Six-Day War, American officials still believed that an American–Soviet arms limitation agreement might be possible. In the early months of 1968, therefore, they sounded out their Soviet counterparts on the issue.[21]

What American officials heard in reply did not please them. The Soviet Union insisted that an arms limitation agreement could only be discussed after Israel had withdrawn unilaterally and completely from the territories that it had captured in the war. The United States responded that the timing and extent of an Israeli withdrawal had to be governed by the terms of United Nations Security Council Resolution (UNSCR) 242 of November 1967, which was supposed to serve as the basis of a comprehensive Arab–Israeli peace settlement. The terms of this resolution, the Johnson administration continued, did not demand that Israel undertake either a unilateral or a complete withdrawal to its pre-Six-Day War borders. By the summer of 1968, to put it differently, the United States recognized that the Soviet Union had no genuine interest in an arms limitation agreement – that its untenable stance represented nothing more than diplomatic cover for its scheme to rearm the Arab states – even as American officials took a significantly less alarmist view of their adversary's conduct than the Eshkol government.

Moreover, the Johnson administration gradually came to the realization that the IAF would not be able to get replacement aircraft from France, including the 50 Mirage Vs that Israel had purchased before the Six-Day War. The IAF's sole source of combat aircraft from the mid-1950s to the mid-1960s, France had slapped Israel with an arms embargo on the eve of the war. Not necessarily permanent at first, it would become so in the months following the war, as France sought to improve its stature in the

Arab world at Israel's expense. France's sale of Mirages to Iraq in early 1968 only increased American pessimism that President Charles de Gaulle's government would reconsider its position on arms sales to Israel.[22]

The Johnson administration also sought to test Israel's readiness to enter into peace negotiations before it made a decision on the Phantom.[23] Specifically, American officials wanted the Eshkol government to participate actively in Arab–Israeli peace talks under the auspices of United Nations envoy Gunnar Jarring.[24] Though they displayed some irritation with the Eshkol government, especially its insistence on direct negotiations with Arab states, its declaration of a united Jerusalem as Israel's capital, and its intentionally vague statements about its willingness to withdraw from territories captured in the Six-Day War, they did detect signs of flexibility in the Israeli diplomatic stance during the course of 1968. Some administration officials even went so far as to claim that the sale of the Phantom, by strengthening Israel's security even further, would probably make the Eshkol government that much more pliable.[25] Indeed, the sale of additional Skyhawks had been predicated in part on the notion that the Eshkol government would take more risks for peace if it received solid evidence of America's commitment to Israel.[26]

Furthermore, American officials reckoned that Egypt, not Israel, constituted the greatest obstacle to the success of the Jarring mission.[27] With the backing of the Soviet Union, President Gamal Abdel Nasser's regime took the position that, under the terms of UNSCR 242, Israel had to withdraw unconditionally and completely from the territories that it had captured in the Six-Day War. Egypt also claimed that negotiations were not necessary to implement the resolution. Its frequent attempts notwithstanding, the Johnson administration, which considered this interpretation of UNSCR 242 to be without merit, could not move Nasser's regime to modify its stand. A Phantom sale to Israel, some American officials believed, might just scare Egypt enough to change its tune.

The American–Jordanian relationship, too, influenced the Johnson administration's decision on the Phantom. The Six-Day War had been a catastrophe for Jordan. All of its territory west of the Jordan River had fallen into Israeli hands. Its army had suffered a severe blow, losing most of its heavy weapons. Because the United States had supplied arms to Jordan before the war, King Hussein's regime naturally asked the Johnson administration to rebuild its army with replacement weapons.[28]

This request placed the United States on the horns of a familiar dilemma. On the one hand, if the Johnson administration rejected the Jordanian request, Hussein's regime would most likely turn to the Soviet Union for arms. Jordan, in fact, had openly threatened to do so, just as it had before the Six-Day War. Soviet arms in Jordan, of course, would mean the loss of this state to the pro-Western camp, a development the

Johnson administration clearly sought to prevent. On the other hand, selling arms to Jordan would call into question the administration's commitment to Israel, and it would elicit howls of public protest from the Eshkol government.

The United States responded to this dilemma in the same way that it had before the Six-Day War. Both Jordan and Israel would get American arms. The additional Skyhawks were sold to Israel with the understanding that the Eshkol government would acquiesce in the sale of replacement Patton tanks (and other equipment) to Jordan.[29] And American officials informed their Jordanian counterparts that the United States could not refrain from providing arms to Israel if it supplied them to Jordan. The Hussein regime and the Eshkol government appeared satisfied with this arms policy.

The same dynamic operated in the case of the Phantom, albeit not quite as explicitly. While the United States did not tie the sale of this aircraft directly to an arms deal with Jordan, the Johnson administration knew that it would have to provide Hussein's regime with more weapons in the future in order to maintain its pro-Western orientation; therefore, Israel had to be supplied with compensatory arms.

The Consummation of the Sale

In light of Soviet and French arms policies, Israeli and Arab attitudes toward the Jarring mission, and American–Jordanian relations, the Johnson administration had decided in principle to sell the Phantom to Israel by the summer of 1968. The precise timing and terms of a sale, however, still had to be negotiated with the Eshkol government. The United States did not intend simply to "give" the Phantom to Israel; the administration wanted the Eshkol government to make certain concessions in return. The nature of these concessions would lead to a series of very contentious bargaining sessions between American and Israeli officials.

The Johnson administration had been worried about Israel's surface-to-surface missile (SSM) and nuclear weapons research programs ever since it entered office. Based on its eagerness to obtain SSMs, American officials inferred that Israel had chosen to develop nuclear weapons, as these missiles would have practically no military value if armed with conventional high-explosive warheads. The scope of Israel's research in the nuclear field itself, much of it carried out at its Dimona reactor facility, strongly reinforced this speculation. While Israel had allowed American inspectors to visit Dimona periodically in order to certify that nuclear weapons were not under construction, the Eshkol government proved notoriously evasive in discussions with American officials regarding its attitude toward these weapons, further heightening the administration's concern.

The Johnson administration had tried earlier to pressure the Eshkol government into renouncing SSMs and nuclear weapons during negotiations to sell the M-48 Patton tank. On this occasion, Israel had fiercely resisted the attempt to link the acquisition of tanks to its SSM and nuclear weapons research programs. The Johnson administration and the Eshkol government eventually settled on a compromise formula. The 10 March 1965 Memorandum of Understanding (MoU), in which the United States committed itself to the sale of tanks to Israel, laid out this formula. The Skyhawk deal cemented this compromise.

Still, the Johnson administration was never really comfortable with this arrangement. It preferred instead that Israel divest itself of all of its SSM and nuclear weapons capabilities. The sale of the Phantom, therefore, provided a convenient forum for the administration to pressure Israel once again on nonconventional weapons. Warnke and Rusk, neither of them known to be particularly understanding of the Israeli predicament, would act as the point men for the DoD and DoS, respectively, in the administration's quest to strip Israel of its nonconventional weapons capabilities.

In mid-October, National Security Council Staff member Harold Saunders wrote to Rostow to inform him that the DoD and DoS would recommend to Clifford and Rusk that the United States connect the sale of the Phantom to Israeli concessions on SSMs and nuclear weapons.[30] Though American officials had urged their Israeli counterparts to forgo ballistic missiles and atomic bombs throughout the year – the Johnson administration showed particular interest in getting Israel to sign the Nuclear Non-Proliferation Treaty (NPT) – they had essentially refrained from linking SSMs and nuclear weapons directly to American arms transfers.[31] For its part, the Eshkol government had responded to even the slightest hint of a connection in these overtures with a demand that the issue of the Jewish state's nonconventional weapons programs be handled separately from the issue of American arms.[32]

Despite the fact that the DoD and DoS knew that the Eshkol government would respond furiously, Rusk linked, however tentatively, Israel's nonconventional weapons programs and a Phantom sale in a late October meeting with Israeli Foreign Minister Abba Eban.[33] The Eshkol government reacted as expected, angrily rejecting any connection. Israel, it asserted, would only be willing to extend to the United States the same sort of assurances that it had given during the Skyhawk negotiations: it would reaffirm its commitment not to be the first state to introduce nuclear weapons into the Middle East and it would promise not to use the Phantom as a nuclear weapons delivery vehicle.[34]

The tension only grew worse once formal negotiations over the sale got underway. The DoD, tasked with leading the talks, wanted the Eshkol government to approve of the following set of conditions in return for the Phantom: (1) Israel would not test, manufacture, or deploy SSMs without the consent of the United States; (2) Israel would not test, manufacture, or deploy nuclear weapons without the consent of the United States; (3) Israel would sign the NPT; and (4) Israel would allow American inspectors ready access to any of its defense installations that the United States deemed it necessary to visit.[35] The Johnson administration had the right to impose these conditions, Warnke told Rabin, because the sale of the Phantom fundamentally altered the American–Israeli relationship. Contrary to its past policy, the United States had now agreed to become Israel's primary source of arms. Warnke mentioned in this breath not only the proposed 50 Phantoms, but also the past sale of 100 Skyhawks, a possible future sale of a second batch of 50 Phantoms, as well as the provision of sophisticated ordnance, jet engines, and other equipment.

The Eshkol government's reply came swiftly. Rabin told Warnke that Israel considered the Johnson administration's conditions to be totally unacceptable.[36] The Eshkol government, he said, would not concede Israel's sovereignty for a mere 50 aircraft. It simply could not go beyond the concessions that it had made previously in connection with the Skyhawk sale. To do otherwise would be to endanger Israel's security.

Just as it seemed that the United States and Israel had reached a serious impasse, Johnson resolved the problem. Despite the advice of Rusk, Clifford, and their staffs, he refused to link a Phantom sale to Israel's SSM and nuclear weapons programs.[37] While he surely wanted the Eshkol government to sign the NPT – he personally requested Israel's compliance with the treaty on at least two occasions – he would not make the sale conditional on the Eshkol government's acceptance.[38] Indeed, he had told Eban as much during their late October meeting, which possibly stiffened the Israeli resolve to withstand DoD–DoS pressure.[39]

Why Johnson supported the Israeli position has never become crystal clear. His well-known fondness for Israel surely accounted for part of his decision. So surely did his belief that Israel had a real need for the Phantom in light of Soviet and French arms policies and Arab intransigence. Also, apparently he alone among senior administration officials had been informed by Central Intelligence Agency Director Richard Helms that Israel already possessed nuclear weapons.[40] If this report is correct, then Johnson may have understood that the Eshkol government would never sign the NPT, no matter how often the administration urged it to do so.

Whatever the truth behind Johnson's thinking, the way had been opened for Israel to obtain the Phantom on its terms. Nevertheless, Warnke made one last effort to link the sale of the Phantom to Israel's nuclear arsenal. The Eshkol government had agreed to reconfirm that

Israel would not be the first state to introduce nuclear weapons into the Middle East as a condition of the sale. Warnke sought to define "introduce" to mean the possession of bomb components, and he further sought to include in the sales contract a clause to the effect that the United States had the right to demand the return of the Phantoms if the Jewish state were ever discovered to be in noncompliance with its terms.[41]

Rabin countered that, in the Eshkol government's eyes, a nuclear weapon that had been neither publicly declared nor field tested had not been brought into the region; Israel, in other words, argued that it could build nuclear weapons without actually introducing them into the Middle East. Moreover, he said that the Eshkol government could not agree to a clause in the sales contract that would allow the United States to take back the Phantoms based upon its meaning of "introduce."[42]

In the end, the Johnson administration and the Eshkol government hammered out a mutually acceptable compromise. They agreed to disagree on the meaning of "introduce," and they agreed that the United States would not seek to cancel the Phantom sale so long as Israel refrained from publicly declaring or field testing a nuclear device. Israel had got the Phantom on its own terms.

5

National Interests or Domestic Politics?

The Rationale Behind American Arms Sales to Israel in the 1960s

The United States faced competing sets of national interests in deciding whether to sell arms to Israel during the 1960s. On the one hand, it had committed itself to ensuring Israel's security, even though it had repeatedly refused to give the Jewish state any sort of formal guarantee. The United States had also committed itself to maintaining a balance of power between Israel and the Arab world. On the other hand, it did not relish the idea of becoming a major arms supplier to the Middle East, and it certainly did not want to get directly embroiled in the Arab–Israeli conflict, particularly at a time when much of its diplomatic efforts and military resources were tied up in the Vietnamese quagmire. Moreover, the United States had important political and economic interests in the Arab world – namely, preventing the spread of Soviet (and radical Arab) influence and protecting the West's oil supplies – and establishing an arms pipeline to Israel could put them in jeopardy. The Johnson administration worried constantly about creating a polarized Middle East, one in which the United States would be identified as Israel's benefactor and the Soviet Union would be identified as the Arab world's benefactor.

A series of developments in the region during the 1960s brought into sharp relief the tension between these competing sets of interests. First, both before and after the 1967 Six-Day War, the Soviet Union poured very large quantities of advanced weapons into the arsenals of the Arab world, a trend that had begun with the Czech–Egyptian arms deal in 1955. The numbers of aircraft, tanks, artillery tubes, naval vessels, and so on in the Egyptian, Syrian, and Iraqi orders of battle climbed steadily upward to unprecedented levels.

Concomitantly, both before and after the Six-Day War, the Soviet

Union brushed aside repeated American attempts to reach an agreement that would limit arms transfers to the region.[1] One American official crisply summarized the Soviet Union's attitude toward arms control in the following manner: ". . . he [Ambassador Anatoly Dobrynin] indicated that his Government considered this [arms control in the Middle East] a very complex question and the tenor of his remarks was to the effect that they [sic] had little interest."[2]

In view of its stance on the Jewish state's security and the Arab–Israeli balance of power, the United States could not be insensitive to the implications of Soviet arms deliveries to the Arab world. Indeed, well before the 1965 sale of M-48 Patton tanks to Israel, the Johnson administration had communicated to Arab governments the message that the United States might eventually have to sell arms to Israel to compensate for imbalances produced by the Soviet weapons entering their arsenals.[3]

This sentiment intensified as Israel's traditional Western European sources of arms dried up by the mid-1960s. Historically, the United States had encouraged the Jewish state to meet its arms requirements through purchases there. Throughout the 1950s and into the 1960s, Israel had been able to procure sufficient quantities of weapons from Great Britain, France, and the Federal Republic of Germany; however, by 1965, the tide had turned heavily against the Jewish state. When the Algerian War ended in 1962, France sought to repair its fractured relationship with the Arab world. Consequently, the Franco–Israeli relationship went into decline, and the Eshkol government found it increasingly difficult to convince France to furnish new arms. Great Britain had never been too keen to supply weapons to the Jewish state because of British ties to the Arab world, and this reticence applied with special vigor to the case of aircraft, like the Buccaneer, that could potentially carry nuclear arms. West Germany supplied arms clandestinely to Israel in the late 1950s and early 1960s, but as soon as the American–German–Israeli tank deal of late 1964 came to light, the Federal Republic immediately cut off all further weapons shipments to the Jewish state.

The Eshkol government, of course, brought intense pressure to bear on the Johnson administration to take over responsibility for filling Israel's arms needs. Even as the United States continued to insist that Israel search in Western Europe for arms, it also recognized the growing magnitude of the Jewish state's plight. The 1965 M-48 tank sale had been motivated in part by Israel's inability to acquire suitable equipment in Western Europe. Deputy Special Assistant to the President for National Security Affairs Robert Komer nicely captured the Johnson administration's feelings just prior to the 1966 A-4 Skyhawk sale:

> But the drying up of Israel's regular European sources . . . *forced us to become*

> *direct suppliers* [italics in original] – first Hawks and then tanks. Since our own deep commitment to Israel's security would almost force us to intervene in another Arab–Israeli flareup [sic], it is in our interest to help Israel maintain a sufficient deterrent edge to warn off [Egyptian President Gamal Abdel] Nasser and other eager beavers.[4]

Similar logic applied with respect to the 1968 sale of the F-4 Phantom. The Eshkol government argued – and the Johnson administration eventually conceded – that Israel required this aircraft as a means to deter its Arab opponents, particularly after France refused to supply the Jewish state with Mirage Vs.

Israel's nuclear research and surface-to-surface missile (SSM) programs pushed the United States in the direction of arms sales, too. The Johnson administration expressed concern that the Jewish state appeared to be on the road toward acquiring both nuclear weapons and long-range SSMs. On several occasions, it tried to halt these programs by linking their termination to American arms sales; however, the Eshkol government had no intention of giving up these programs in the absence of a formal American security guarantee, especially while Egypt retained its weapons of mass destruction.

If it could not stop Israel's nuclear and SSM programs, the administration decided, the next best alternative would be to exercise a degree of control over them. Arms transfers, it realized, could be used as leverage. The United States would supply Israel with weapons in exchange for the Eshkol government's pledge to keep its nuclear and SSM programs largely hidden in the shadows. The March 1965 Memorandum of Understanding on M-48s incorporated the agreed upon formula in Clause II: "The Government of Israel has reaffirmed that Israel will not be the first to introduce nuclear weapons into the Arab–Israeli area."[5] Thus, the Israeli policy of nuclear "opacity" was born.[6] Israel would not publicly flaunt its nuclear and SSM programs.

The terms of the Skyhawk sale leave no doubt that the United States intended the deal in part to reinforce Israel's pledge on these programs. Once again, Komer neatly summed up the administration's position:

> *Can we use planes as a lever to keep Israel from going nuclear?* [italics in original] Desperation is what would most likely drive Israel to this choice, should it come to feel that the conventional balance was turning against it. So judicious US arms supply, aimed at maintaining a deterrent balance, is as good an inhibitor as we've got.[7]

And, the Phantom sale was driven by precisely the same logic insofar as concerned nuclear arms and SSMs. The United States, in short, could not compel Israel to forsake forever nuclear weapons and SSMs, but it could make them less appealing to the Jewish state by furnishing adequate quantities of conventional arms. Furthermore, periodic American inspections of Dimona would permit the United States to save face, at home and abroad, on the important issue of nuclear nonproliferation.

The Johnson administration also arrived at the conclusion that arms sales to Israel would provide it with a measure of leverage over the Eshkol government's decisions to use military power to protect the Jewish state's national interests. In the mid-1960s, the United States feared that border skirmishing over Arab efforts to divert Israel's water resources and Arab terrorist infiltration of Israel's frontiers could escalate into a full-scale war. The Johnson administration repeatedly insisted that the Eshkol government hold its fire in the face of Arab provocations.[8] Sharp rebukes from the United States on those occasions when Israel resorted to military means to fight water theft and terrorism, however, did not elicit the desired long-term response from the Eshkol government.

While the United States knew that Israel would never forfeit its right to defend its national interests through military means when necessary, the Johnson administration felt that the Eshkol government could be induced to show greater restraint if American arms were forthcoming. In connection with the M-48 sale, Rusk stated:

> It is important to preserve an adequate Arab/Israel [arms] balance, in order to prevent such periodic crises as that approaching over the Jordan waters from flaring into open war which we would have to intervene [in] to stop. To deny arms to Israel will greatly reinforce its tendency to take early preemptive action against the Arabs.[9]

And, in connection with the Skyhawk sale, Komer asserted: ". . . the more secure Israel feels, the less likely it is to strike first, as at Suez. So maintaining a reasonable Arab/Israeli arms balance helps limit the chances of our [the United States] being drawn into a Near East crisis."[10]

Indeed, threats to shut down an arms pipeline could serve to rein in the Eshkol government's propensity to resort to military means. In connection with the M-48 sale, Rusk put it bluntly:

> At [the] same meeting [between American Ambassador to Israel Walworth Barbour and Eshkol] . . . you should tactfully inform Primin that while we are pleased at the successful conclusion of the tank arrangements, he should be aware, however, that we continue [to] adhere to our position against the use of force by Israel against its neighbors. Military strikes by Israel against its neighbors would force us [to] reconsider proceeding with deliveries.[11]

Later in the same communication, he went on to remark that: "By threatening to make continued deliveries contingent on no preemptive strikes we hope [to] continue [to] keep Israel in line."[12]

Arms sale to Israel were also linked to the American–Jordanian relationship. In the mid-1960s, Jordan faced intense pressure from Egypt to accept Soviet arms with which to confront the Jewish state. King Hussein let the Johnson administration know that, if American weapons were not forthcoming, Jordan would have no choice but to enter into an arms relationship with the Soviet Union. The administration, naturally, did not want to let a friendly Arab state drift into the Soviet orbit; therefore, the United States had decided by 1965 that it must sell weapons to Jordan in order to prevent King Hussein from arranging for the delivery of Soviet arms.

The United States, though, could not afford to supply weapons to Jordan without also providing them to Israel. Such a choice, after all, would call into question the dual American commitments to the Jewish state's security and to an Arab–Israeli balance of power. It would elicit vehement protests from the Eshkol government, too. Consequently, the Johnson administration resolved that Israel would get arms as well, particularly as high-ranking officials believed that sales were inevitable in the long run. Rusk and McNamara presented the linkage between arms for Jordan and Israel in connection with the Skyhawk sale in the following way:

> We [the United States] have been under considerable pressure from Israel to sell U.S. aircraft regardless of what we may do for Jordan. If we sell to Jordan there would have to be a compensating sale to Israel. Without the sale to Jordan we might be able to stall the Israelis for a time. However, particularly if we believe that eventually we must provide some aircraft to Israel, a sale now to Jordan would have a number of advantages. It would protect the considerable U.S. investment in that country, would enable us to exert continued influence for stability in the area and, by preempting the Soviets in Jordan, would prevent a major step towards an East–West polarization of arms supply to the area. . . . Our agreement to sell U.S. [aircraft] to Jordan, to which we are convinced there is no feasible alternative, could be incorporated in the same kind of Israel–Jordan package arrangement developed for the ground equipment sales last March.[13]

This dual sales policy actually served American national interests quite well, as the administration fortified two pro-Western states in the face of Soviet (and radical Arab) expansionism in the Middle East.

Finally, the conduct of pro-Soviet Arab states, especially Egypt, had an impact on American arms sales to Israel. The Johnson administration never had any illusions about Egypt's support for the Soviet agenda in the Middle East, which made it easier to enter into an arms relationship with the Jewish state. Moreover, Egypt's unreasonable stand on a negotiated settlement of the Arab–Israeli conflict in the wake of the Six-Day War – its complete inflexibility with respect to United Nations Security Council Resolution 242, coupled with the Eshkol government's perceived flexibility on the matter – helped to convince the administration to sell the

Phantom to the Jewish state. Not only would the sale of this aircraft further encourage Israel's inclination to work toward peace by reassuring it of American support, but it might also encourage Egypt and its Soviet patron to adopt a more sensible position on an Arab–Israeli settlement by illustrating to them that the United States would not let a regional client down.

To this last reason for creating an arms pipeline to Israel might be added the fact that the Johnson administration also ultimately realized that weapons transfers to the Jewish state did not translate into harm to American national interests in the Arab world. Komer summarized the administration's perspective prior to the Skyhawk sale:

> *Will selling planes to Israel spook the Arabs?* [italics in original] Our Arab experts so warned before we sold [Hawk missiles] in 1962, then again before we sold tanks. But actual experience shows far less reaction than we feared. . . . [I]t's a good bet that quiet, limited aircraft supply to Israel will not upset our [the United States] applecart in the Arab world.[14]

No Arab state would be happy about American arms sales to Israel; nevertheless, those states already in the Western camp would not move into the Soviet orbit as a result. Nor could the United States "lose" Arab states already in the Soviet orbit.

The (Minor) Role of Domestic Politics in Arms Sales to Israel

It should be clear by now that a compelling set of reasons related to American national interests crystallized during the 1960s, resulting in the establishment of an arms pipeline from the United States to Israel; however, this set of reasons does not exclude the possibility that domestic politics may also have influenced the Johnson administration's agenda.

Unquestionably, Johnson himself had a friendly attitude toward Israel. Furthermore, he developed very cordial relationships with both Eshkol and Ephraim Evron, the number two man at the Israeli embassy in the United States.[15] Additionally, Johnson, a very astute politician, knew that arms sales to Israel could only help his and the Democratic Party's stature at home.

Still, no real evidence exists to suggest that purely domestic political considerations had a significant impact on American decision making with regard to arms sales. The Johnson administration did not sell M-48s to Israel in order to curry favor with Jewish voters or the pro-Israeli lobby. If it had been worried about appeasing these groups, it would have shipped arms directly to Israel in 1964, an election year. The administration's apparent lack of concern with regard to a domestic backlash over a refusal to provide tanks during this year is attested to by the position of the National Security Council Standing Group on arms to the Jewish state.

> The only foreseeable adverse result would be increased Israeli pressure on the American Jewish community to support the tank request. [This pressure] can be counteracted by a careful explanation of 1) past and present U.S. economic and military assistance to Israel, 2) Israel's present strong military posture, 3) Israel's remarkably flourishing economy, and 4) the nature and extent of U.S. assurances of support for Israel in such matters as security and the Jordan waters off-take.[16]

Only Myer Feldman among important administration officials backed a tank sale to Israel in 1964.

Domestic political considerations, at most, buttressed a sale based on a firm strategic rationale. Officials like Rusk and McNamara would not have endorsed the deal unless they felt certain that it would advance American national interests in the Middle East. Domestic politics, in short, entered into the decision primarily to the extent that the administration sought the Eshkol government's assistance in convincing the Jewish state's American supporters that a substantial arms sale to Jordan would be in the national interests of both the United States and Israel.

The same thinking flowed into the Skyhawk sale. Domestic political considerations did not enter into the administration's decision making beyond the need to get American supporters of Israel behind the simultaneous sale of aircraft to Jordan. Rusk and McNamara noted that "an offer to sell [aircraft] to Israel would make a sale to Jordan more acceptable to circles in which criticism of such a sale could be expected."[17]

While the M-48 and Skyhawk sales occurred mainly behind closed doors, far away from the glare of publicity, the Phantom deal received much comment in the United States throughout 1968.[18] The American Jewish community, particularly in the form of the pro-Israeli lobby, of course, expressed itself very strongly in favor of the sale. Influential national organizations – like Americans for Democratic Action, the American Legion, and the American Federation of Labor–Congress of Industrial Organizations – also approved of the sale. Congress, too, backed the sale; several resolutions that endorsed the dispatch of the Phantom to Israel circulated in the House and Senate during the summer, attracting widespread support among representatives and senators alike. Hubert Humphrey and Richard Nixon, the Democratic and Republican presidential candidates, respectively, both committed themselves to the sale as well.

The Johnson administration's awareness of this groundswell of support, however, does not mean that domestic political considerations discernibly affected its decision to sell the Phantom.[19] Discussions among administration officials on the pros and cons of a deal over the course of 1968 do not indicate that they took any real account of domestic political considerations. Rather, it seems that domestic politics simply gave the administration a useful pretext to justify the sale to an angry Arab world. Rusk even instructed American embassies throughout the Middle East to employ the domestic politics card at their discretion to defuse hostile Arab reactions to the sale.[20] And Johnson himself claimed to have lobbied Humphrey and Nixon to come out in favor of the Phantom deal, perhaps to provide his administration with that much more cover.[21]

Ironically, domestic political considerations actually impeded arms sales to Israel for a brief period after the Six-Day War. The Johnson administration kept the postwar arms freeze in place an extra few months in order to put pressure on Israel to convince its supporters in the United States, especially in the Congress, to permit renewed weapons sales to pro-Western Arab states. Only after the Jewish state agreed to do the administration's bidding, and only after the Congress moderated its attitude toward renewed American–Arab arms deals, did the United States lift the weapons embargo against Israel.

A flight of three Mirage IIICs somewhere over Israel. Perhaps more than any other weapons system, the Mirage symbolized the Franco-Israeli arms partnership from the mid-1950s to the mid-1960s. Photographer: Moshe Milner.

The Israel Defense Forces relied heavily on the British-supplied Centurion tank in both the 1967 Six-Day War and the 1973 Yom Kippur War. Photographer: Eitan Haris.

The United States rescinded its ban on supplying major arms systems to Israel when the Kennedy administration sold several batteries of Hawk anti-aircraft missiles to the Jewish state in 1962. Photographer: Moshe Milner.

The first major arms deal between the United States and Israel during the Johnson administration involved the sale of the M-48 Patton tank. Photographer: Moshe Milner.

An A-4 Skyhawk taking off from an air base in Israel. When it entered Israeli service in late 1967, the Skyhawk added a significant punch to the Israel Air Force's ground attack capabilities. Photographer: Moshe Milner.

F-4 Phantoms flying in tight formation over Israel during an air display. The acquisition of this aircraft allowed the Israel Air Force to conduct punishing "deep-penetration" raids against the Egyptian hinterland during the 1969–1970 War of Attrition. Photographer: Moshe Milner.

The F-15 Eagle has been a mainstay of the Israel Air Force since the mid-1970s. Photographer: Ya'acov Sa'ar.

The Reagan administration temporarily embargoed the delivery of F-16 Fighting Falcons to Israel after they were employed to destroy Iraq's Osirak nuclear reactor in 1981. Photographer: Moshe Milner.

PART II

How the Arms Relationship Has Operated

6
The 1967 Six-Day War

A Delayed "Green Light" for Preemption

By the mid-1960s, the security-for-autonomy bargain at the heart of the American–Israeli patron–client relationship had become operational. Though the United States had not yet pledged itself to becoming Israel's principal supplier of arms, it had already concluded three major arms deals with the Jewish state: the 1962 sale of Hawk anti-aircraft missile batteries, the 1965 sale of M-48 Patton tanks, and the 1966 sale of A-4 Skyhawk aircraft. The United States, in short, had significantly upgraded its commitment to Israel's security during these years.

While the United States had a number of reasons for transferring arms to the Jewish state, Israel's nuclear weapons and surface-to-surface missile (SSM) programs, as well as its penchant for employing military means to protect its national interests, were prominent on the list. The United States did not agree to strengthen Israel at no diplomatic cost to the latter. Rather, in exchange for arms, the United States expected the Jewish state to surrender a measure of control over its nuclear weapons and SSMs as well as to be sensitive to American desires when making decisions about whether to employ the Israel Defense Forces (IDF) to protect the state's interests.

Both the United States and Israel displayed grudging approval about the terms of the security-for-autonomy bargain. The 1967 Six-Day War – the first real test of the bargain – would demonstrate conclusively that it could actually function under wartime conditions. And this war would institute a pattern of conduct on the part of patron and client alike that would be repeated in future conflagrations.

The Prewar Crisis

The third full-scale round of fighting between Israel and the Arab world, the Six-Day War ultimately grew out of border skirmishing between the

Jewish state and Syria in April 1967.[1] The Syrian Air Force's embarrassing loss of six aircraft in one notably intense mêlée temporarily brought matters to a boil between Israel and Syria. Both states, however, backed away from further escalation toward war, and the clashes between them petered out. Still, they continued to exchange overt warnings and threats in the following weeks.

Always keen to stir up trouble in the Middle East in a quest to expand its influence in the region, the Soviet Union took advantage of the tension generated by the Israeli–Syrian clashes to accuse the Jewish state of concentrating the IDF along the border in preparation for a large-scale assault on its northern Arab neighbor. Even though Israel immediately and strongly repudiated this phony allegation, even offering to permit foreign observers to tour the area to validate the fact that the IDF had not been mobilized for an attack, the Soviet charge led to dramatic developments that sent the Middle East hurtling toward war.

In mid-May, cajoled by fellow Arab states and the Soviet Union to "defend" Syria against Israeli "aggression," Egypt demanded the removal of United Nations (UN) peacekeeping troops, who had been stationed along the Egyptian–Israeli border since the end of the 1956 Sinai Campaign, as part of the agreement to secure the IDF's withdrawal from Gaza and the Sinai in the wake of its lightning victory in that war. Egypt also began to reinforce heavily its military assets in the Sinai, a process that would continue up to the outbreak of war. Not to confront Israel at this juncture, Egypt reasoned, would threaten its position as leader of the Arab world.

Israel, of course, mobilized the IDF in response to Egypt's military moves in the Sinai. Initially, the IDF deployed in a defensive posture, simply to counter the possibility of an Egyptian attack into the Negev Desert or into the vulnerable coastal plain. The former contained Israel's only port on the Red Sea, through which moved its maritime trade with the Far East, while the latter contained most of the state's population centers and industrial facilities. But, soon after the IDF had secured the state's borders, it moved to positions that would allow for a swift attack into Egypt.

The United States responded with caution to what had suddenly blossomed into a major Arab–Israeli showdown. During the initial phase of the crisis, which lasted for about a week, the Johnson administration sought to defuse the situation before events spiraled out of control. The administration did not formulate a specific plan to end the crisis, but it did hope to restore the *status quo ante*. It wanted UN peacekeeping troops sent back to their posts along the Egyptian–Israeli border, and it wanted both Egypt and Israel to demobilize their armies.

To accomplish these objectives, the Johnson administration purposely asked the Jewish state to give diplomacy time to bring the crisis to a peaceful resolution. President Lyndon Johnson himself urged Israel in a less-than-subtle way not to make a hasty decision to resort to military means when he wrote to Prime Minister Levi Eshkol.

> I know that you and your people are having your patience tried to the limits by continuing incidents along your border. In this situation, I would like to emphasize in the strongest terms the need to avoid any action on your side which would add further to the violence and tension in your area. I urge the closest consultation between you and your principal friends [i.e., the Western powers]. I am sure that you will understand that I cannot accept any responsibilities on behalf of the United States for situations that arise as the result of actions on which we are not consulted.[2]

If Israel initiated hostilities without prior American consent, the Johnson administration had signaled, the Jewish state would find itself on its own.

The Eshkol government agreed to the American request to accede to diplomacy, even though it felt that the United States had not taken a strong enough stand on Israel's behalf. Indeed, Eshkol wrote to Johnson:

> Your note . . . does not explicitly refer to the commitment by the United States to act both inside and outside the UN in support of Israel's integrity and independence. I understand that you do not wish to be committed without consultation. But with a massive build-up [sic] on our southern frontier . . . and Soviet support of the governments responsible for the tension, there is surely an urgent need to reaffirm the American commitment to Israel's security with a view to its implementation should the need arise.[3]

The Eshkol government wanted a forthright declaration that the United States would back the Jewish state in the crisis.

Israel had two principal reasons for accommodating the United States on the matter of diplomacy. First, the Jewish state had been caught by surprise by the outbreak of the crisis, because its military intelligence service had predicted that the next Arab–Israeli downward spiral toward war would not occur before the 1970s. The IDF not only required a breathing period in which to mobilize, deploy, and prepare its forces for combat, but it also needed time to revise its operational plan to fit the current military circumstances. Second, and equally important, the Eshkol government did not want the Johnson administration to be in a position to say that Israel had not granted the United States an opportunity to end the crisis without bloodshed and, therefore, that the latter had no obligation toward the former. Foreign Minister Abba Eban expressed this mindset when he wrote that the Eshkol government did not want the Johnson administration to "claim that Israel had not involved it [the United States] frankly in its [Israel's] dilemma."[4]

Despite efforts to defuse the situation, the tension between Israel and the Arab world grew more ominous. During the intermediate stage of the crisis, which also lasted for about a week, Egypt acted on its previously declared threat to shut the Straits of Tiran to merchant vessels destined for Israel, thereby blockading the port of Eilat. Consequently, Egypt entered into a formal state of war with the Jewish state under the rules of international law. And it continued to augment its military forces in the Sinai. Moreover, the Arab world intensified its war of nerves with Israel by openly calling for the Jewish state's total and violent eradication.

The Johnson administration now had three goals in the crisis: to ensure that UN peacekeeping troops returned to their positions along the Egyptian–Israeli border, to ensure that Egypt and Israel demobilized their armies, and to ensure the lifting of the blockade of the Jewish state's southern port. The administration placed special emphasis on the third objective, and began to consider the idea of putting together an international naval flotilla to guarantee the principle of freedom of the seas.

The United States kept up the pressure on Israel to show restraint in the face of Egyptian provocations. Even before Egypt announced its decision to interdict shipping to and from Israel, Johnson wrote forthrightly to Eshkol:

> Ambassador [Walworth] Barbour has informed me of your assurances . . . that the measures your Government is taking are precautionary in nature and that you will continue to do all you can to avoid further deterioration of the present grave situation on your borders. By continuing to display steady nerves you can, I am convinced, make a major contribution to the avoidance of hostilities.[5]

This pressure to demonstrate restraint did not abate in the immediate aftermath of Egypt's decision.

Even as the Johnson administration assured the Eshkol government that the United States would respond with understanding to Israel's plight, American officials insisted that the Jewish state had to give the United States more time to achieve both a national and international consensus on how to end the crisis. During a visit by Eban to Washington to discuss the situation, Secretary of State Dean Rusk, Secretary of Defense Robert McNamara, and Johnson all urged the Eshkol government in no uncertain terms to refrain from any sort of preemptive military strike. Indeed, Johnson wrote to Eshkol: "As your friend, I repeat even more strongly what I said yesterday to Mr. Eban. Israel just must not take any preemptive military action and thereby make itself responsible for the initiation of hostilities."[6]

The mood in Israel darkened measurably during this stage of the crisis. Many of the generals in the IDF's high command argued that the Jewish state ought to attack Egypt right away rather than allow the wheels of diplomacy to grind on. The longer Israel held its fire, these officers asserted, the more death and destruction it would suffer in a war that they had come to see as inevitable, especially in light of the fact that the state's very right to exist had been publicly challenged by the Arab world.

Some members of Eshkol's cabinet essentially concurred with this dreary assessment of the situation, pointing out the considerable economic and psychological damage that indefinite mobilization and diplomatic paralysis would inflict on the state and its citizenry. Nevertheless, in deference to American wishes, the Eshkol government chose to refrain for the moment from a military strike to allow more time to achieve a diplomatic settlement that would protect Israel's interests. According to Eban, "Eshkol now asked the Cabinet to give the United States . . . a chance for a 'few more days' to succeed – or to acknowledge failure – in [its] efforts. Of the eighteen ministers, only one . . . was prepared to vote for immediate military action."[7] Eshkol himself explained the decision as follows:

> Had we not received Johnson's letter and Rusk's message, I would have urged the Government to make the decision to fight; but their communications pointed out not only that unilateral Israel [sic] action would be catastrophic but also that the United States was continuing with its preparations for multilateral action to open the Gulf to shipping of all nations. I could not forget that the letter was signed by the President who had once promised me face-to-face: "We will carry out whatever I ever promise you." I did not want him to come afterwards and say, "I warned you in advance and now you cannot make any claims whatever on the United States and its allies."[8]

The Israeli decision to grant the United States more time to find a peaceful settlement to the crisis notwithstanding, the situation continued to spiral toward war. During the third, and final, stage of the crisis, which lasted less than a week, Jordan entered into a war coalition with Egypt and Syria, placing its military forces under the command of the Egyptian army. Iraq moved a significant portion of its military forces into Jordanian territory, itself a cause for war from the Israeli perspective. Smaller military contingents from other Arab countries deployed for combat alongside the Egyptian army. The Palestinians, too, made preparations for war. Arab rhetoric against Israel became ever more hysterical. War seemed all but certain to break out at any moment.

The Johnson administration's efforts to reach a peaceful settlement went on, but the administration itself essentially recognized during this stage of the crisis that war had now become inevitable. Arab and Soviet intransigence, combined with Western and Third World indifference, had doomed all efforts, including the one to put together an international naval flotilla to end the blockade of the Straits of Tiran. Once the United States realized that a peaceful solution was not in the offing, it washed its hands of the crisis. The administration refused to consider any unilateral American action to resolve the situation.

Concomitantly, however, it signaled to the Eshkol government that it would no longer be opposed to Israeli military action. Contrary to the initial and intermediate stages of the crisis, when the Johnson administration urged restraint on the Eshkol government, the final stage of the crisis was marked by a relative absence of similar demands. The United States, in short, gave the Jewish state a tacit "green light" to embark on war.

The Eshkol government received this signal loudly and clearly, especially after MOSSAD (Israel's foreign intelligence service) director Meir Amit returned from an early June visit to Washington. He emerged from his conversations with senior American officials convinced that, while the United States would do nothing to help Israel prosecute the war, it would understand and support the Jewish state's decision to fight, a message that he expressed to the Israeli cabinet.[9] Eban echoed this view:

> We had gained enormously by patiently allowing the United States to test its plan for international action up to its total collapse. As I examined the [diplomatic] cables, I observed that we were now being released from the weight of Johnson's pressure. Secretary of State Dean Rusk had been asked by the press whether any efforts were being made to keep Israel from precipitate action. His reply had been: "I don't think it's our business to restrain anyone." . . . I believed that this [military action] would now be received with unspoken relief . . . in Washington.[10]

On 2 June, Defense Minister Moshe Dayan, who had joined the government just a day earlier, approved the IDF's operational plan to attack the Egyptian army. On 4 June, aware that no help would be forthcoming from abroad, that Israel's very existence was in jeopardy, and that the United States would not abandon the Jewish state, the Eshkol government opted for war. On the morning of 5 June, the IDF struck hard at Egypt in the air and on the ground.

Analysis of American and Israeli Conduct

American national interests drove the conduct of the Johnson administration during the prewar crisis. The United States believed that, if war did occur, Israel would defeat its Arab opponents rather swiftly.[11] An Israeli victory, the administration thought, would lead to a sharp deterioration in the relationship between the United States and the Arab world, because the latter would blame the former for its setback. Moreover, Soviet influence in states like Egypt, Syria, and Iraq, all of which had already aligned themselves with the Eastern bloc, would expand in the wake of an Israeli victory, posing a heightened threat to pro-Western, oil-producing Arab states like Saudi Arabia and Kuwait.

The Johnson administration, naturally, sought to avoid these expected negative consequences, which also included the possibility of an interruption of oil supplies and a loss of access to military facilities in the region.[12] On the other hand, given the basic American commitment to Israel's survival, as well as the administration's fear that the abandonment of a client state would gravely damage American prestige around the globe, the United States refused to contemplate cutting the Jewish state loose in its hour of need. With national interests at stake in both the Arab world and Israel, therefore, the administration wanted to restore peacefully the *status quo ante*. This solution, according to its way of thinking, would protect American national interests in both the Arab world and the Jewish state. To this end, the Johnson administration applied strong pressure on Israel to refrain from military action so long as it appeared that the *status quo ante* could be restored through diplomatic means.

Once the administration realized that the crisis would not be resolved peacefully, however, it tacitly approved of the Eshkol government's decision to go to war, because it would not allow the United States to become embroiled in a face-to-face confrontation with the Arab world that could result in an armed clash between American and Arab military forces. An Israeli victory in war, the Johnson administration concluded, would do less harm to American–Arab relations than would an American–Arab confrontation. Moreover, the administration reasoned that the United States could not afford to involve itself in another regional conflict at a time when it was bogged down in Vietnam.[13] The Johnson administration, in short, glumly accepted that an Arab–Israeli war would undermine the regional *status quo* to the disadvantage of the United States; but it also believed that the new postwar *status quo* would be far less unfavorable to Washington if the United States were to avoid direct participation in hostilities.

Israeli perceptions of the importance of the American–Israeli relationship drove the conduct of the Eshkol government during the prewar crisis. Though the government did not expect that the Jewish state would receive American arms during the war – it thought that the duration of the fighting would be too short to permit any sort of foreign military assistance – it nevertheless did not intend to take the risk that the United States might be alienated by its decision to initiate hostilities; therefore, the implicit consent of the United States had to be secured before the IDF could be unleashed against Israel's foes.

Based on the IDF's calculations, the Eshkol government had no doubt that the Jewish state would defeat its Arab opponents in a war. But it also believed that the cost of victory could be quite high in death and destruction. Furthermore, the government had to assume that an Israeli triumph would not lead to a resolution of the Arab–Israeli conflict. Israel would have to replace its losses in order to prepare itself for future rounds of warfare. It would also need postwar support in the diplomatic arena to ensure that it would not be stripped of the fruits of victory, as it had been to a large extent in the aftermath of the Sinai Campaign. Finally, the Eshkol government had at least to think about the potential for Soviet intervention in the fighting on behalf of the Arab world, however remote this prospect might appear.

If the United States were to deter the Soviet Union during the war and to provide Israel with arms and diplomatic backing after the war, then the Eshkol government had to ensure that the Jewish state remained in America's good graces. This logic explains the government's decision to delay military action until such time as the Johnson administration had recognized that diplomacy had run its course.

Israel had kept its part of the security-for-autonomy bargain during the prewar crisis, so the United States kept its part of the bargain.[14] During the war, the Johnson administration made sure that the Soviet Union did not intervene on behalf of the Arab world. After the war, it did not compel Israel to surrender the fruits of the IDF's victory. United Nations Security Council Resolution 242, which had strong American sponsorship, contended that the Jewish state did not have to return any of the territory that it had captured in the Six-Day War until the Arab world made peace with it and implied that some of this territory could be retained on a permanent basis. Most importantly, following an interruption lasting several months, the United States renewed arms shipments to Israel and eventually became its principal supplier of weapons.

7

The 1969–1970 War of Attrition

Restricting Israel's Military Options

The existence of the security-for-autonomy bargain notwithstanding, the United States and Israel continued to hold very different views on a number of contentious Middle Eastern issues after the 1967 Six-Day War, not least in regard to their particular visions of how to achieve a final settlement of the Arab–Israeli conflict. The gap between their positions would become plainly evident during the next Arab–Israeli war, the 1969–1970 War of Attrition. This gap would, consequently, strongly influence how the security-for-autonomy bargain functioned during the later stages of the war.

Escalation and Deescalation During the War

The Six-Day War had ended in a dramatic military victory for Israel. The Egyptian, Syrian, and Jordanian armies had been thrashed to the point of utter ruin. Moreover, Israel's geographical situation had been dramatically altered for the better by the hostilities. To the north, it had captured the Golan from Syria. To the east, it had seized Judea and Samaria from Jordan. And, to the south, it had taken the Gaza Strip and the Sinai from Egypt. The Jewish state had attained defensible borders and a degree of strategic depth for the first time in its brief history. Its population centers, industrial assets, and military facilities no longer remained within easy reach of Arab armies.

Still, despite its military victory, Israel did not receive any respite from war. Indeed, the conquest of the Sinai had set the stage for what would come to be known as the War of Attrition, as Egypt preferred to regain its territory through fighting rather than through talking.[1] Egypt made its stance on a diplomatic settlement with the Jewish state quite clear at the Khartoum Conference of 1967 in the form of the three "noes" – no direct negotiations with Israel, no peace with Israel, and no recognition of Israel.

For its part, the Jewish state had adopted a more flexible attitude in the wake of the Six-Day War. It displayed a certain willingness to hand back the Sinai to Egypt, perhaps with minor, security-related border adjustments; however, it would only agree to do so in exchange for direct negotiations that would result in a peace treaty that acknowledged once and for all its right to exist within secure and recognized frontiers. The Egyptian and Israeli positions, in short, were simply irreconcilable.

Not long after the end of the Six-Day War, therefore, Egypt decided to heat up the Egyptian–Israeli frontier, which now ran along the Suez Canal. The Egyptians believed that, through massive artillery bombardments, occasional air strikes, and sporadic infantry raids, the Egyptian army would be able to inflict an unacceptable level of casualties on the Israel Defense Forces (IDF). Furthermore, renewed hostilities between Egypt and Israel would inevitably draw the United States and the Soviet Union into the Arab–Israeli conflict to the Jewish state's detriment. A combination of Egyptian military pressure and superpower diplomatic pressure, according to Egyptian calculations, would eventually compel the Jewish state to abandon the Sinai without extracting any concessions from Egypt in return.

Once the Soviet Union had restocked and rebuilt its army to a sufficient level, Egypt began the implementation of its plan to recapture the Sinai. For almost two years, from the summer of 1967 to the spring of 1969, the Egyptian–Israeli border saw intermittent, but often quite serious, fighting that followed the same general pattern. Egypt would open fire on Israeli targets along the Suez Canal in an effort to cause heavy casualties. The Jewish state would respond either with attacks on Egyptian targets along the canal or deep in the rear. In late 1968, for example, when the Egyptian army initiated a large-scale bombardment against Israeli targets, the IDF responded by striking infrastructure facilities, such as bridges and power transformers, in the Nile Valley.[2] On other occasions, it implicitly threatened even more sensitive facilities by conducting reconnaissance flights over them.[3] The Jewish state's actions would calm the border for a period of time; however, eventually, Egypt would resume the fighting, eliciting another round of Israeli counterblows. This pattern of sporadic, tit-for-tat combat dragged on until March 1969.

During that month, Egypt officially repudiated the cease-fire that had brought the Six-Day War to a close and officially proclaimed the beginning of the War of Attrition.[4] The first stage of hostilities, very similar in nature to the fighting that had occurred over the previous two years, though much heavier and far more sustained, lasted from March until mid-July. Egypt had decided to intensify hostilities because its earlier military efforts had neither loosened Israel's grip on the Sinai nor enticed the United States and the Soviet Union to involve themselves in the fighting.

Likewise, Egypt's strategy during the first stage of the war did not induce the Jewish state to withdraw from the Sinai, nor did it prompt the superpowers to intervene in the hostilities; however, it did inflict a substantial number of casualties on the IDF. Prime Minister Golda Meir's government, therefore, decided to escalate the war in mid-July by committing the Israel Air Force (IAF) to combat on a grand scale. The government reasoned that a massive air assault against Egyptian military targets along the Suez Canal would serve to "thin out" their ranks, thereby leading to a concomitant reduction in IDF losses, and would also deter Egypt from undertaking the large-scale, cross-canal offensive that had been promised at an opportune moment. Israel, in sum, chose to escalate the war in an attempt to bring it to a speedy conclusion on its terms. This second stage of the war, during which the IAF destroyed Egypt's entire air defense network in the canal zone, would last until January 1970.

Until the second stage of the War of Attrition, the United States had essentially kept its distance from the renewed Israeli–Egyptian fighting. It had engaged in periodic talks with the other members of the "big four" – the Soviet Union, Great Britain, and France – both inside and outside of the United Nations (UN) to examine the prospect of finding a comprehensive solution to the Arab–Israeli conflict, including the canal-side hostilities; but these talks had not been pressed with any genuine sense of urgency and ultimately amounted to nothing. The United States had also consented to provide Israel with additional arms, most notably F-4 Phantom aircraft, especially after the Soviet Union had shown no inclination to curb arms supplies to its Arab clients. Neither the administration of President Lyndon Johnson nor the following administration of President Richard Nixon, though, saw any pressing reason as of yet to inject the United States into the Arab–Israeli conflict, particularly because the Jewish state had more than held its own in the fighting to date.

The Nixon administration's perspective underwent a rapid revision as the War of Attrition grew more intense. In December 1969, it issued the "Rogers Plan," named after Secretary of State William Rogers, its chief proponent, which envisaged a comprehensive solution of the Arab–Israeli conflict. The administration's initiative foundered almost immediately, as Israel, Egypt, and the Soviet Union each objected strenuously to different parts of the plan, which suffered an unheralded death not long afterward. Nevertheless, the release of the Rogers Plan clearly signaled a new American activism *vis-à-vis* the Arab–Israeli conflict.

In January 1970, the War of Attrition entered its third stage. Though the IAF had wreaked tremendous damage on Egypt's military forces in the vicinity of the Suez Canal during the second stage of the fighting – and though it had prevented Egypt from mounting a large-scale, cross-canal operation to recover the Sinai – Israel had not been able to bring the war itself to a close. Egyptian artillery in particular continued to pound Israeli targets, inflicting significant numbers of casualties on the IDF.

The desire to stem further losses, coupled with its fear that, after the release of the Rogers Plan, the United States and the Soviet Union might possibly impose a solution of the Arab–Israeli conflict on their respective clients, one that could be unfavorable to the Jewish state's national interests, convinced the Meir government that the war had to be brought to a swift conclusion. In an effort to do so, consequently, it decided to escalate the fighting once more.

From January to April, the IAF carried out a series of "deep-penetration" air raids against military targets in the Egyptian hinterland. On 13 January, to cite one example, IAF aircraft attacked the military camps at Khanka, a mere 12.5 miles northeast of Cairo. On 28 January, to cite another example, they hit the military camps at Ma'adi, just over 6 miles south of Cairo. The IAF conducted approximately three dozen deep-penetration raids during these months, sometimes bombing the same target on several different occasions.[5]

From a narrow military standpoint, the deep-penetration strikes proved to be rather effective. They caused a significant amount of destruction to their targets, and they shook Egyptian morale. From a broader strategic perspective, on the other hand, they did not serve their intended purpose. Not only did they completely fail to bring an end to the war, but they also had two other very unwelcome results for Israel.

First, they triggered direct, large-scale Soviet involvement in the fighting to prevent an Egyptian defeat in the war. At the behest of Egypt, the Soviet Union deployed thousands of troops to serve in the field alongside the Egyptian army. By April, their numbers had reached the range of 6,500–8,000. By June, their ranks may have swelled to as many as 12,000. Soviet troops manned numerous anti-aircraft missile batteries and piloted 120 aircraft.[6] And, as events would show, they were not noticeably shy about engaging in combat with the IAF.

Second, and even more worrisome, the Nixon administration did not react to the deep-penetration raids in the manner anticipated by the Meir government. The Jewish state's ambassador in the United States, Yitzhak Rabin, had urged the government to undertake the raids apparently secure in the belief that the Nixon administration would back them. Israeli Foreign Minister Abba Eban observed that:

> [In late December 1969] Rabin returned to Israel from his post in Washington ostensibly to discuss recent American proposals for a peace settlement with Egypt [i.e., the Rogers Plan] but in fact to take part in the crucial discussion of the question of in-depth bombing of Egypt. His strong views in favor of this course were already well known from his telegrams, some of which were formu-

lated in terms of sharp rebuke to his government for failing to understand what he termed "an irrevocable opportunity." We in the cabinet had a deep interest in hearing his views on the potential response of the United States to an escalation of our pressure against Egypt.[7]

While Eban personally may not have been swayed by Rabin's logic, the Meir government echoed the ambassador's thinking.

But, contrary to expectations, the Nixon administration voiced increasing opposition to the deep-penetration raids as the months rolled on.[8] It eventually gave its displeasure concrete expression in March by refusing to follow through on an Israeli petition to purchase additional Phantom and A-4 Skyhawk aircraft. According to Rogers, "the President has decided to hold in abeyance for now a decision with respect to Israel's request for additional aircraft," ostensibly because the IAF had a sufficient order of battle to protect the Jewish state.[9] In the face of evident American resistance, a stunned Meir government determined that the deep-penetration raids had to be brought to a halt.

The termination of these raids ushered in the War of Attrition's final stage, which lasted until an early August cease-fire agreement put a stop to the fighting. During this stage, Israel fought a steadily faltering battle to retain its air superiority in the Suez Canal zone. Soviet and Egyptian troops systematically pushed forward a new anti-aircraft defense network. Despite suffering heavy losses in men and equipment in the process, they exacted a substantial price from the IAF in downed aircraft, especially in precious Phantoms. Israel's most notable victory in these months occurred in late July, when IAF aircraft destroyed five Soviet-piloted MiGs in a classic dogfight.

The Nixon administration, alarmed by the prospect of widening hostilities, especially in the form of a full-blown Soviet–Israeli conflagration, now sought to achieve an Egyptian–Israeli cease-fire. To gain the Jewish state's "cooperation," the administration adopted a "carrot-and-stick" approach.[10] On the one hand, the United States indicated that its arms pipeline to Israel would essentially remain shut unless the Meir government consented to a cease-fire. But, on the other hand, the administration promised to provide the Jewish state with arms and diplomatic support if it accepted the American initiative.

Despite some hesitation at first, Israel eventually accepted the American cease-fire proposal. A combination of American pressure and promises, Eban remarked, constituted a decisive element in the Jewish state's decision making.

In late July, when we [the Meir government] surveyed our situation in all its aspects, we felt that the risks of accepting the American cease-fire proposal were far less than the dangers of rejecting it. Rejection would mean the continuation

> of savage war with Egypt, the prospect of involvement in military conflict with the Soviet Union, and diminishing American fidelity to Israel.[11]

He went on to observe the benefits that Israel would derive in return for its acquiescence to American foreign policy. In a presidential message to the Meir government:

> President Nixon promised 1) that there would be a continuation and even a reinforcement of American economic and military aid, even if there were a cease-fire, which might ostensibly justify the diminution of that aid; 2) Israel would not be required to withdraw its forces from the occupied territories until a satisfactory peace treaty had been concluded. Until that time, said President Nixon, Israel would not be called upon by the United States to remove a single Israeli soldier from existing positions; 3) Israel would not be required to accept a solution of the [Arab] refugee problem that might alter the Jewish character of the state; and 4) the U.S. would protect us [Israel] in the UN Security Council against peremptory orders to sacrifice our existing territorial positions without adequate and credible security or political gains.[12]

These pledges, the Meir government knew, went beyond any previous American commitment to Israel.

In the short term, the United States committed itself to supply highly sophisticated electronic countermeasures that could deflect Soviet anti-aircraft missiles from their targets as well as advanced precision-guided munitions that would make air attacks on anti-aircraft missile launchers and radar sets more effective. Additionally, the Nixon administration hinted that, farther down the road, Israel could expect to receive considerable numbers of additional Phantoms and Skyhawks, particularly if the balance of power shifted against the Jewish state in the event that Soviet–Egyptian forces were to violate the cease-fire agreement.

The Meir government had initially balked at the American cease-fire proposal precisely because it feared that Soviet–Egyptian forces would violate the agreement in order to gain a tactical advantage over Israel by strengthening their air defense network in the canal zone. This concern proved to be prescient. Not too long after the cease-fire took effect, IDF reconnaissance assets spotted massive Soviet–Egyptian violations of the agreement in the form of new anti-aircraft missile batteries located closer to the Suez Canal. The Meir government, of course, demanded that the United States enforce the cease-fire agreement and return the situation along the canal to the *status quo ante*.

For its part, the Nixon administration refused even to acknowledge the existence of any violations for a few weeks. When it finally did so, it had harsh words for the Soviet Union and Egypt; however, it took no tangible action to reverse the violations, nor would it countenance any Israeli moves to do so. Instead, the United States decided to compensate the Jewish state with a huge arms package that included additional Phantoms and Skyhawks, as well as tanks, artillery tubes, and other equipment.[13]

The Meir government complained strenuously of the seeming American indifference to the Soviet–Egyptian cease-fire violations. Nevertheless, in the absence of American backing, Israel did not seriously contemplate the resumption of hostilities. Instead, it accepted the offer of arms, thereby once again surrendering a measure of its independence in exchange for American assistance.

Analysis of American and Israeli Conduct

The desire to prevent a deterioration in the post-Six-Day War *status quo* in the Middle East drove the conduct of the United States during the War of Attrition. The Nixon administration believed that the United States had three basic interests in the region. First, it sought to contain the spread of radical Arab and Soviet influence, mainly in order to protect pro-Western Arab states, particularly their oil reserves. Second, it sought to avoid any direct military involvement in the Arab–Israeli conflict at a time when American military forces remained bogged down in Vietnam. Even more important, it sought to steer clear of the possibility that its military forces could become embroiled in combat with Arab or Soviet military forces, as the development of such a scenario could lead to dire consequences, including a full-scale war between the superpowers. And, third, it sought to make certain that the balance of power did not shift against Israel.

Until it formulated the Rogers Plan, the United States largely contented itself with supplying arms to Israel, partly in order to offset the steady stream of Soviet arms flowing into radical Arab states, especially Egypt. These arms ensured that the balance of power would not shift against the Jewish state and meant that the United States would not be dragged into Middle Eastern hostilities in defense of Israel. Neither the Johnson nor Nixon administration attached a significant price tag in terms of Israeli concessions with respect to the Jewish state's ongoing conflict with Egypt so long as the fighting did not seem to threaten unduly American interests in the region.

The intensification of the War of Attrition, and the burgeoning prospect of direct Soviet intervention on behalf of Egypt, however, changed the Nixon administration's thinking about the Middle East. The Rogers Plan, which envisioned major Israeli concessions as part of a comprehensive solution to the Arab–Israeli conflict, constituted an American effort to prevent a deterioration of the post-Six-Day War *status quo*. Indeed, the administration reasoned that American influence in the region might grow

– and Soviet influence might decline – if the United States could broker a comprehensive solution to the Arab–Israeli conflict.

When Israel's deep-penetration raids triggered direct Soviet intervention in the War of Attrition, the United States promptly blamed the Jewish state for creating the conditions that led to the expansion of Soviet influence in the region and to the prospect that the superpowers might eventually come to blows on behalf of their respective clients. To counter these developments, which the Nixon administration obviously viewed as inimical to American national interests, the United States put pressure on Israel to halt the deep-penetration raids by conditioning future military and diplomatic assistance on their termination.

For the same reasons, the Nixon administration used a similar combination of threats and inducements in regard to future military and diplomatic assistance to get the Jewish state to agree to a cease-fire and then to acquiesce in Soviet–Egyptian violations of the cease-fire. An end to hostilities, the administration reasoned, would stem the spread of Soviet influence in the Middle East, defuse the prospect of a superpower confrontation, assure the Jewish state's security, and could serve as the basis of renewed Arab–Israeli peace negotiations, which might eventually redound to the benefit of the United States.

The conduct of Israel during the War of Attrition was highly sensitive to fluctuations in the American–Israeli patron–client relationship. So long as the United States did not insist on specific Israeli actions in exchange for arms transfers and diplomatic backing, the Meir government could follow its own instincts about how to prosecute the War of Attrition. Thus, because the Nixon administration did not object, the Jewish state could escalate the fighting in July 1969 in an attempt to bring the war to a close on its terms.

Based on the lack of a negative American response to past Israeli actions, as well as its firm belief that the Nixon administration would not be unhappy to see Egypt – the greatest threat to pro-Western Arab states – thrashed in the War of Attrition, the Meir government initiated the deep-penetration bombing campaign in January 1970.[14] When the United States did not react as expected to the bombing raids, and would not provide the military aid and diplomatic support necessary for the raids to go on, Israel felt that it had no choice but to stop them. Concern over Soviet intervention, to be sure, affected the Meir government's decision making on this matter; however, the Soviet role should not be overestimated in Israeli thinking, as the Jewish state displayed no hesitation in taking on Soviet forces in battle later in the war. The deep-penetration raids most likely would have continued had the Nixon administration sanctioned them.

Likewise, the Meir government consented to a cease-fire and then acquiesced in Soviet–Egyptian violations of the cease-fire, because it calculated that the IAF could not fight the combined forces of the Soviet Union and Egypt with any long-term prospect of victory without a firm American commitment to provide the required military assistance and diplomatic support, particularly in light of growing Israeli aircraft losses during the last stage of the War of Attrition. The Jewish state, in short, had to bring the war to a close on the Nixon administration's terms, because of its dependence on the United States. This reality would lay the foundations for the next Arab–Israeli conflagration, the 1973 Yom Kippur War.

8

The 1973 Yom Kippur War

Limiting Israel's Military Victory

The American–Israeli patron–client relationship experienced a period of unusual tranquility from fall 1970–fall 1973, principally because, in American eyes, the Arab–Israeli conflict took a back seat to radical Arab and Soviet troublemaking in the Middle East during these years. The Nixon administration even began to regard the Jewish state as a "strategic asset" that, under certain circumstances, could serve as a bulwark against radical Arab and Soviet expansionism in the region, especially after the Israel Defense Forces (IDF) had assisted in the survival of Jordan's pro-Western regime in September 1970. Consequently, the United States transferred significant quantities of arms to Israel without attaching any real strings. This rosy (for the Jewish state) situation came to an abrupt halt with the outbreak of the 1973 Yom Kippur War.

American Arms Transfers and the War

"No war, no peace" aptly summarized the condition of the Arab–Israeli conflict after the 1969–1970 War of Attrition. The Arab world rejected the postwar *status quo*, which saw the Jewish state still in complete control of the territories that it had captured in the 1967 Six-Day War. Unable to wrest these lands away from Israel by force, it remained unwilling to pay the price – peace treaties with the Jewish state – that would have been necessary to recover them peacefully. Israel, to the contrary, approved of the postwar *status quo*, and therefore showed little inclination to part with territory in exchange for anything less than peace treaties with its neighbors. All diplomatic efforts to break this stalemate, including the major push of 1971, came to naught.

Consequently, Egypt and Syria concluded that another round of full-scale warfare would be necessary for them to achieve a resolution to the Arab–Israeli conflict consistent with their national interests. These states believed that, if their armies could inflict heavy losses on the IDF, seize and hold pieces of the Sinai and Golan, respectively, and involve the United States and the Soviet Union in the fighting, then they could fulfill their strategic aims, even if they were ultimately to suffer defeat on the battlefield.

Egypt and Syria still required the military means to implement their design. They recognized that they would neither be able to exact a significant toll in men and machines nor capture territory unless they could counter the IDF's overwhelming air and armored superiority. Massive Soviet arms transfers in the form of the most modern anti-aircraft and anti-tank guided weapons available at the time, however, gave them these means. By fall 1973, then, Egypt and Syria had a viable war option.

Despite the fact that Israel's military and civilian intelligence services had gathered copious amounts of high quality data about Arab arms acquisitions, operational plans, troop movements, and so forth, both the IDF and Prime Minister Golda Meir's government remained convinced until the day war actually erupted that the Jewish state did not face an imminent prospect of hostilities.[1] The problem lay not in the realm of information gathering, but rather in the sphere of information processing.

The Arab world, especially Egypt, had spoken on more than one occasion in the early 1970s about engaging in war "to restore Arab rights," but had not acted on its words. Rhetoric of this sort unaccompanied by any action served to lull both the IDF and the Meir government into a sense of complacency. The Arab world, they believed, would not challenge the Jewish state on the battlefield so long as Israel remained the strongest power in the region. They did not conceive of the notion that the Arab world might be willing to endure a military setback in order to achieve a diplomatic triumph.

Blinded by this assumption, both the IDF and the Meir government misinterpreted fresh information in the days before the outbreak of war that pointed in the direction of hostilities. When Egypt and Syria mobilized their armies and deployed them along the cease-fire lines in the Sinai and on the Golan, Israel considered this development to be nothing more than routine military exercises or, perhaps, saber rattling (in response to a humiliating Syrian air defeat in mid-September in which 13 MiGs had been shot down in a dogfight). And, when Soviet citizens departed Egypt in large numbers, the Jewish state chalked this development up to a feud between the Soviet Union and its client.

Only at the last moment, literally hours before the beginning of the Arab assault, when they received irrefutable proof of Egyptian and Syrian intentions from an unimpeachable source, did the IDF and the Meir government acknowledge their terrible mistake. Nevertheless, Israel still had sufficient time to prepare and execute a preemptive air strike at the outset of the Yom Kippur War.[2] The General Staff sought immediate authorization for a preemptive strike, arguing that, though it would not be as devastating as the Jewish state's opening attack during the Six-Day War, such a strike would put an unready IDF, which needed a minimum of 48 hours to mobilize and deploy its reserve formations, the core of its combat power, in a far better position to withstand the initial Arab onslaught.

The Meir government, though, refused to sanction a preemptive attack, a decision that the Nixon administration, particularly Secretary of State Henry Kissinger, encouraged in an emphatic manner.[3] Indeed, the secretary of state had long cautioned Israel against a preemptive strike in conversations with the Jewish state's officials. Israeli Ambassador to the United States Simcha Dinitz remarked that "Dr. Kissinger had always told me, whatever happens, don't be the one that strikes first. He told this to [former Israeli Ambassador to the United States Yitzhak] Rabin too."[4] And, on 6 October, the day the war began, Kissinger sent a cable to President Richard Nixon saying that:

> We are urgently communicating with the Israelis, warning them against any preemptive action. . . . I then called Israeli Chargé [d'Affaires Mordechai] Shalev [and] I emphasized to him the essentiality of restraint on the Israeli part, and said there must be no preemptive strike. . . . Shalev called back shortly thereafter and said his government assured us there would be no preemptive action. Shortly thereafter we received a message from Prime Minister Meir confirming this.[5]

Caught between the IDF's General Staff and the Nixon administration, the Meir government chose to follow the position of Israel's patron rather than the advice of its own military experts.

Rather than potentially wresting the initiative away from the Egyptian and Syrian armies, therefore, the IDF had to absorb the first blow, resulting in serious reverses during the early days of the fighting. Not only did it suffer heavy losses among men and machines, but it also gave ground in the Sinai and on the Golan. By the second day of the war, the IDF appeared to be in dire straits on both fronts.

From the moment the war broke out, Israel asked the United States to provide arms. While the Nixon administration approved the dispatch of small amounts of weapons to be carried in Israeli cargo aircraft within a couple of days, it dragged its feet on the issue of a large-scale American supply effort.[6] Not until 14 October did the first American cargo aircraft arrive in the Jewish state with arms for the IDF as part of an emergency airlift operation. Almost a week had gone by since a similar Soviet airlift of arms to its Arab clients had got underway. Moreover, the Nixon administration did not approve the airlift until after Egypt had categorically rejected an American attempt to arrange a cease-fire.

Even though American arms were not forthcoming at the start of the Yom Kippur War, the qualitative superiority of Israeli forces *vis-à-vis* their Arab opponents soon began to manifest itself on the battlefield. In the north, the IDF smashed the Syrian army, drove it from the Golan, and even managed to seize a sizable chunk of additional territory before the first American cargo aircraft landed on Israeli soil. In the south, it halted the Egyptian offensive into the Sinai, defeating a major armored thrust toward the peninsula's vital passes, and counterattacked across the Suez Canal into Egypt proper, all before additional American weapons started to reach its forces in meaningful quantities (though the promise of those arms may have affected to some degree the pace, scope, and intensity of the counteroffensive).

With the IDF crossing of the Suez Canal into Egypt, the Yom Kippur War moved into its final stage. In contrast to the opening week of the war, Egypt now actively sought an end to the fighting in order to avert the total collapse of its armed forces. The United States and the Soviet Union tried to hammer out a durable cease-fire; however, a series of attempts to stop the hostilities proved abortive, as both Egyptian and IDF units continued to fight in an effort to improve their respective military positions.

The IDF promptly took advantage of Egyptian cease-fire violations to move deeper into Egypt, eventually surrounding the Egyptian Third Army. The Nixon administration quietly approved of the IDF's advances in the face of Egyptian violations.[7] But it would not countenance the destruction of the Third Army itself.[8] Despite getting into a dangerous nuclear showdown with the Soviet Union, as both superpowers sought to back their clients in public, the United States placed great pressure on Israel to spare the Egyptian army. The Nixon administration even suggested to the Meir government that the United States would stand aside while the Soviet army joined the fighting to save the Third Army.[9]

Israeli Foreign Minister Abba Eban phrased the Jewish state's dilemma this way: "Should we attempt the destruction of Egypt's Third Army at the risk of Soviet intervention, or should we ensure American support . . . by allowing the Third Army to be saved?"[10] After some haggling with the Nixon administration over the precise terms of a cease-fire agreement, the Meir government ultimately chose to loosen the IDF's grip on the Third Army, thereby reducing the magnitude of Israel's victory in the Yom Kippur War.

Disagreement between the United States and Israel did not come to an end when the guns fell silent. During postwar disengagement negotiations, the Nixon and, later, Ford administrations placed considerable pressure on the Jewish state to make concessions to Egypt and Syria, especially in the form of withdrawals from portions of the Sinai and Golan, respectively.[11] To get Israel to make the desired concessions, the United States once again used a carrot-and-stick approach.

On the one hand, the Nixon and Ford administrations promised to bolster the American–Israeli relationship, including the arms pipeline, if the Jewish state were to give back chunks of the Sinai and Golan. On the other hand, implicitly at times and explicitly at others, the United States threatened to punish Israel, holding out the possibility of a disruption in arms transfers, if the Jewish state were to reject American demands. Indeed, the Ford administration even initiated a "reassessment" of the entire American–Israeli relationship, implying that the United States could drastically reduce its military, economic, and diplomatic support of Israel if the Jewish state did not satisfy American wishes.

For its part, Israel resisted American pressure for as long as it could do so without actually harming the bilateral relationship; however, in the end, it gave in to American demands. It eventually withdrew from a large portion of the Sinai and a small part of the Golan in exchange for increased American backing.

Analysis of American and Israeli Conduct

The conduct of the United States in the Yom Kippur War grew out of the same set of national interests that had driven American policy during the Six-Day War and the War of Attrition. First, the United States sought to minimize radical Arab and Soviet influence in the Middle East, primarily in order to protect pro-Western Arab states and their oil resources. Second, the United States did not want to get into a conflict in the region, because its armed forces still remained bogged down in Vietnam. Furthermore, a military confrontation with the Arab world or the Soviet Union could lead to the direst consequences for the United States, including the possibility of a full-scale war between the superpowers. And, third, the United States sought to make sure that Israel's basic security was not put in jeopardy.

In contrast to the Six-Day War and the War of Attrition, however, when the United States sought merely to prevent a deterioration in the regional *status quo*, the Nixon administration spied a golden opportunity in the Yom Kippur War to alter the *status quo* to American advantage. Both Nixon and Kissinger concluded that the United States could significantly roll back Soviet influence, and greatly expand American influence, in the Middle East, particularly in Egypt, if the administration played its cards right during the war.

A regional realignment, though, would not occur if Israel were to win an overwhelming military victory like it had in the Six-Day War. The Nixon administration, therefore, sought to engineer a battlefield stalemate, particularly between the Jewish state and Egypt, in order to promote postwar negotiations, which the United States would then mediate to its own benefit. Nixon's words clearly reveal the administration's intention:

> I believed that only a battlefield stalemate would provide the foundation in which fruitful negotiations might begin: Any equilibrium – even if only an equilibrium of mutual exhaustion – would make it easier to reach an enforceable settlement. Therefore, I was convinced that we must not use our influence to bring about a cease-fire that would leave the parties in such imbalance that negotiations for a permanent settlement would never begin.[12]

Israel's dependence on American arms would be manipulated by the administration in an attempt to achieve this objective.

Before the shooting began, the United States firmly believed that the IDF would swiftly crush its Arab opponents on the battlefield. Hence, the Nixon administration strongly discouraged an Israeli preemptive attack, implying that the Jewish state should not expect to receive so much as a bullet from the United States if it opened hostilities, because such a strike would only add to the IDF's ability to smash its Arab foes. The same calculus led the United States to drag its feet for over a week in dispatching arms to the Jewish state. Not until after it had become evident that the IDF had suffered some early reverses on the battlefield, not until after a massive Soviet airlift had been under way for days, and not until after Egypt had rejected the notion of a cease-fire did the United States authorize a matching airlift to Israel. Kissinger summarized the American position as follows:

> We pursued this [a cease-fire] until Saturday of the first week – that is to say until 13 October. On 13 October it was clear that the Soviets could not deliver the Egyptians to what was in effect a cease-fire in place, and to which we had obtained Israeli acquiescence, more or less. When that occurred we felt we had no choice except to go another route. . . And this is the reason why we started the airlift. . . . It is the principal reason why we started the airlift. . . . [13]

The United States, to put it another way, only began an airlift once its first attempt to engineer a battlefield stalemate had wrecked itself on the shoals of Egyptian intransigence, leaving the Jewish state at a potential military disadvantage in the face of continuing Soviet arms shipments to the Arab world.

Egypt's refusal to agree to a cease-fire early in the war meant that the Nixon administration could not in the end bring about the battlefield stalemate that it envisioned originally; however, the United States could still limit the scope of the Israeli triumph in the war by using arms deliveries to gain leverage over the Jewish state. Nixon himself made this very point:

> In order to have the influence we need to bring Israel to a settlement, we have to have their [sic] confidence. That is why [the] airlift. . . . [W]e have to do enough to have a bargaining position to bring Israel kicking and screaming to the table.[14]

Thus, the United States permitted Israel to surround the Egyptian Third Army (in order to improve America's bargaining position *vis-à-vis* Egypt and the Soviet Union), but it would not allow the Jewish state to destroy this army (in order to acquire Egyptian favor). The Nixon administration, in short, spared Arab "honor," making it possible for the United States to mediate postwar negotiations without meaningful Soviet participation. This diplomatic agenda also accounts for the administration's carrot-and-stick approach to Israel during postwar negotiations.

The Jewish state's conduct in the Yom Kippur War, like its conduct in the Six-Day War and the War of Attrition, is not comprehensible unless it is examined in the context of the American–Israeli patron–client relationship. Unquestionably, the Meir government's fear of a negative American reaction constituted the decisive reason why it dismissed the IDF's advice to launch a preemptive attack. The Meir government believed that, if Israel struck the initial blow, the United States would not assist the Jewish state during the war.[15] Defense Minister Moshe Dayan summed up his government's attitude: "if American help was to be sought, then the United States had to be given full proof that it was not we [the Israelis] who desired war – even if this ruled out preemptive action and handicapped us in the military campaign."[16] Had the Meir government been certain of American support in the wake of a preemptive strike, it would almost certainly have consented to such an attack; however, faced with a powerful Arab war coalition that had the unshakeable support of the Soviet Union, Israel simply could not risk the loss of American goodwill.

Likewise, the Jewish state's dependence on American arms convinced the Meir government to accept, albeit tentatively, a Nixon administration cease-fire proposal during the first week of the war. Similarly, the administration's threat to abandon Israel compelled the Meir government to loosen the IDF's grip on the Egyptian Third Army later in the war. Indeed, in order to explain this decision, Meir herself made explicit reference to the American–Israeli relationship:

> There is only one country to which we can turn and sometimes we have to give in to it – even when we know we shouldn't. But it is the only real friend we have, and a very powerful one. We don't have to say yes to everything, but let's call things by their proper name. There is nothing to be ashamed of when a small country like Israel, in this situation, has to give in sometimes to the United States. And when we do say yes, let's for God's sake not pretend that it is otherwise . . . .[17]

Finally, the Jewish state felt that, in the aftermath of a very destructive war, one in which its internal capabilities had been heavily drained by the fighting, it had no alternative but to trade the postwar concessions desired by the United States for continued American support, particularly in the area of arms transfers. A firm American–Israeli patron–client relationship, the Jewish state reasoned, took precedence over the retention of chunks of the Sinai and Golan in regard to its national interests.

9

Peacetime Arms Transfers

The Nixon, Carter, and Reagan Administrations

In wartime, the security-for-autonomy bargain has operated aggressively, as illustrated by developments in the American–Israeli relationship during the 1967 Six-Day War, the 1969–1970 War of Attrition, and the 1973 Yom Kippur War. During these wars, a very close connection often existed between American arms, on the one hand, and Israeli foreign policy concessions, on the other hand. In peacetime, to the contrary, this bargain has functioned in a more subtle fashion, as attested to by developments in the American–Israeli relationship during the Nixon (1969–1974), Carter (1977–1980), and first Reagan (1981–1984) administrations. The connection between American arms and Israeli foreign policy concessions proved to be more tenuous at times when the Jewish state was not embroiled in a full-scale war.

The Nixon Administration

The strains that afflicted the American–Israeli relationship in the War of Attrition and the Yom Kippur War were absent during the interwar years.[1] Indeed, not only did President Richard Nixon's administration and Prime Minister Golda Meir's government avoid serious differences over policy from fall 1970 to fall 1973, but American and Israeli national interests also converged to an extraordinary degree, particularly with the outbreak of conflict in Jordan.

Not long after the War of Attrition drew to a close, the Palestine Liberation Organization (PLO) fomented a civil war in Jordan.[2] It sought to overthrow the pro-Western regime of King Hussein in order to spur the resumption of fighting between Israel and its Arab neighbors. With the backing of the Soviet Union, Syria decided to intervene in the civil war on behalf of the Palestinian insurgents; however, demonstrative maneuvers in the air and on the ground by the Israel Defense Forces (IDF) deterred the

Syrian army from deploying large-scale forces to assist the Palestinians. Consequently, Jordan's Arab Legion was able to eject the Syrian invaders and crush the Palestinian insurgents.

The United States had encouraged Israel's passive support of Hussein's regime throughout the civil war. Hence, in late 1970, the Nixon administration, partly to reward Israel for its actions during the Jordanian crisis, extended a $500 million line of credit to the Jewish state to purchase arms, including additional F-Phantom and A-4 Skyhawk aircraft, tanks, armored personnel carriers, artillery tubes, and missiles.[3]

Other reasons entered into the Nixon administration's calculus as well. First, the United States wanted to ensure that Israel continued to hold its fire against Egypt, particularly after the Soviets and Egyptians had violated the terms of the cease-fire agreement that concluded the War of Attrition. Second, the Nixon administration sought to encourage greater Israeli cooperation in peace negotiations then taking place under the auspices of United Nations (UN) envoy Gunnar Jarring. Finally, the United States might also have wanted to reinforce the tacit agreement between the two states whereby Israel would keep its nuclear weapons and ballistic missile programs under wraps.[4]

Though the Meir government decided to resume participation in UN-mediated peace talks at the Nixon's administration's insistence, Israel did not noticeably soften its position in these negotiations. Nor did the Jewish state display significantly greater flexibility during later talks under American auspices to arrange an interim settlement between Israel and Egypt along the Suez Canal. Neither of these peace-making efforts eventually yielded any tangible results in resolving the Arab–Israeli conflict.

Yet, the Jewish state's tough diplomatic stance had only a slight impact on American arms transfers in the years leading up to the Yom Kippur War. After a delay of several months, perhaps to apply pressure to the Meir government, the Nixon administration agreed to a follow on deal for more Phantom and Skyhawk aircraft in the waning days of 1971 in exchange for nothing more concrete than Israeli consent to participate in American-sponsored negotiations.[5] The United States sold another batch of Phantoms and Skyhawks to the Jewish state in early 1973, this time around without even making a pretense of linking the sale to Israeli diplomatic concessions. And, in 1972–1973, Israel acquired additional tanks, armored personnel carriers, artillery tubes, missiles, and electronic equipment from the United States, again without any diplomatic concessions on its part.[6]

The Carter Administration

An air of tension hung over the American–Israeli relationship throughout most of the Carter administration. Much of the discord emanated from differences of opinion between President Jimmy Carter and Prime Minister Menachem Begin on three distinct issues: the direction of the Arab–Israeli peace process; the scope of aircraft sales to Israel, Egypt, and Saudi Arabia; and the intensity of Israeli counterinsurgency operations in Lebanon.[7]

The Carter administration entered office with an approach to Arab–Israeli peacemaking that differed significantly from its predecessors. Whereas the Nixon and Ford administrations, under the guidance of Secretary of State Henry Kissinger, had opted for a "step-by-step" model of peacemaking, the Carter administration initially intended to achieve a "comprehensive" peace. That is, contrary to the Nixon and Ford administrations, which sought to make incremental headway in the peace process by tackling each discrete issue before moving on to the next one, the Carter administration preferred to deal with all of the outstanding problems between Arabs and Israelis at one fell swoop.

Consequently, the administration early on revived the concept of a Geneva Peace Conference – a concept that had been defunct since the immediate aftermath of the Yom Kippur War. At a projected summit, an Arab delegation, including Palestinian representatives, and an Israeli delegation would negotiate a comprehensive peace agreement under the joint auspices of the United States and the Soviet Union, with the superpowers potentially stepping in to mandate solutions to any problems that the parties could not work out between themselves.

The Begin government responded quite negatively at first to the notion of a renewed Geneva Peace Conference. It rejected the idea of Palestinian representation, fearing that such a presence would be used to undermine its claims to Judea, Samaria, and Gaza. Of even greater importance, the Begin government objected to joint American–Soviet chairmanship of a conference, believing that this formula constituted a prescription for an imposed solution to the Arab–Israeli conflict – one that would not be favorable to Israeli national interests. Eventually, though, it reluctantly accepted both the concept of an international peace conference as well as a Palestinian presence at such a gathering (so long as the representatives had no formal affiliation with the PLO).

After the Begin government appeared to soften its attitude toward an international peace conference as well as toward Palestinian representation at that gathering, the Carter administration finally approved the sale of attack helicopters to Israel.[8] The timing of this particular sale, especially in light of the fact that the deal had been delayed in the past, indicates that the administration had chosen to reward the Jewish state for its "good" conduct.

Like its predecessors, the Carter administration adopted a carrot-and-stick approach to arms sales to Israel. During its first year in office, it reneged on the Ford administration's promise to supply cluster bombs to the Jewish state and it refused to sanction the coproduction of the F-16 Fighting Falcon aircraft in Israel. In contrast to these negative decisions, the administration allowed some other weapons sales to go forward, and it permitted Israel to invest about $100 million of American aid money in the development of its own Merkava tank. In each case, the administration's calculus was motivated in part by a desire to influence Israeli conduct on issues related to the peace process, such as Jewish settlements in Judea, Samaria, and Gaza.[9]

Furthermore, during the long negotiations that finally culminated in the Israeli–Egyptian peace treaty of March 1979, administration officials suggested at points during the talks when the Begin government appeared to be rather inflexible, particularly over issues such as Palestinian autonomy, that the United States might delay (or even withhold) the supply of specific arms to the Jewish state. Concomitantly, administration officials let it be known that Israeli flexibility would result in substantial arms transfers. Indeed, once the Begin government relaxed its position on Palestinian autonomy, turning the Israeli–Egyptian peace treaty into a reality, not only did the United States provide the Jewish state with a large infusion of new weapons, including an expedited delivery of F-16s previously on order, but it also agreed to build new airfields for the Israel Air Force (IAF) in the Negev to replace those bases in the Sinai that would be lost when the area reverted to Egyptian control.[10]

In February 1978, months before Israel and Egypt began serious peace talks, the Carter administration announced that it wanted to sell 75 F-16s and 15 F-15 Eagle aircraft to the Jewish state, 50 F-5E Freedom Fighters to Egypt, and 60 F-15s to Saudi Arabia.[11] Publicly, the administration presented the sale as a means both to strengthen the national security of three pivotal pro-American states and to promote peaceful coexistence in the Middle East. Secretary of State Cyrus Vance said that:

> We [the Carter administration] have concluded . . . that the sales of these aircraft to the countries in question will help to meet their legitimate security requirements, will not alter the basic military balance in the region, and will be consistent with the overriding objective of a just and lasting peace.[12]

When asked whether the administration also intended to send a message to Israel to be more flexible in peace talks by selling aircraft to Egypt and Saudi Arabia, Carter quickly dismissed the suggestion.[13] But his National Security Adviser, Zbigniew Brzezinski, later acknowledged that the administration sought in part to "punish" the Begin government for its alleged intransigence.[14]

Like Brzezinski, the Begin government rejected Carter's claim. It noted that, under the terms of the proposed sale, Israel would only receive about half of the number of F-16s that it had been promised by the Ford administration. Moreover, the Begin government could not help but notice that the Carter administration, by offering F-15s to Saudi Arabia, had now put that state on par with Israel in regard to the quality of arms supplied by the United States. To the Israeli way of thinking, these aspects of the deal were meant to send an unspoken message to the Begin government to change its tune on peacemaking or suffer a further erosion in administration backing.

The implied American démarche notwithstanding, the Begin government opposed the aircraft sale, because, in its opinion, Israel's national security would be harmed as a consequence. Begin himself observed that:

> Israel cannot . . . under any circumstances agree to link the supply of planes promised it with the supply of offensive planes to Egypt, or of very advanced planes to Saudi Arabia. With all due respect, then, I call upon the President of the United States to reconsider this decision . . . for it constitutes a grave danger . . . to Israel's security.[15]

Though an increase in the number of F-15s to be sold to Israel, along with administration assurances that the Saudi Arabian F-15s would be stripped of ancillary equipment that could turn them into a threat to the Jewish state, did nothing to lessen the Begin government's hostility to the deal, these amendments proved sufficient to overcome domestic opposition to the sale; therefore, the deal received formal approval in May 1978.

The Carter administration and Begin government also knocked heads repeatedly over Israeli counterinsurgency operations in Lebanon.[16] After an especially deadly terrorist attack that killed over 30 civilians, the IDF launched a large-scale incursion into southern Lebanon in March 1978 to clear PLO guerillas from the Jewish state's northern border. Not only did the Carter administration criticize the IDF operation itself and demand an Israeli withdrawal to the international border, but it also hinted that arms transfers to the Jewish state might be affected if it were determined that American weapons had been employed in violation of end-use restrictions. Under American pressure, the Begin government soon withdrew the IDF to the international border, though it left a friendly Lebanese militia, not UN peacekeeping troops, in control of the Lebanese side of the frontier.

Relatively calm for awhile, skirmishes between the Jewish state and the PLO along the Israeli–Lebanese border picked up in 1979–1980. The Carter administration once more objected to Israeli counterinsurgency operations in southern Lebanon. Predictably, the administration again raised the issue of end-use restrictions on American arms, and it again hinted that weapons transfers to Israel might be affected if the Jewish state were found to have employed arms improperly. Under American pressure once more, the Begin government again curtailed IDF operations in southern Lebanon.

The Reagan Administration

The American–Israeli relationship during the first Reagan administration exuded an air of warmth generally absent during the Carter years, perhaps because the United States now considered the Jewish state an important linchpin in the formation of an anti-Soviet "strategic consensus" in the Middle East.[17] Nevertheless, the relationship went through some rocky moments, and arms transfers sometimes figured into these bilateral disagreements. The same issues that bedeviled the American–Israeli relationship during the Carter administration – the Jewish state's policy in the administered territories, its opposition to arms sales to Arab states, and its military operations – riled the waters during the Reagan era.

In contrast to the Carter years, a major American–Israeli disagreement over the administered territories that involved arms transfers erupted not about the status of Judea, Samaria, and Gaza, but rather about the status of the Golan.[18] In December 1981, the Begin government, perhaps in an effort to undermine domestic critics of Israel's impending final withdrawal from the Sinai under the terms of the Israeli–Egyptian peace treaty, extended Israeli law to the Golan, a step that actually fell short of formal annexation.

Nevertheless, the United States objected strenuously to the Begin government's move, postponing the implementation of a strategic cooperation agreement signed a couple of weeks earlier as well as interrupting some sizable arms transactions. In the words of Secretary of State Alexander Haig:

> . . . the President decided that the cost would be a suspension of the U.S.–Israeli agreement on a Memorandum of Understanding (MoU) establishing limited strategic cooperation between the two countries, which had been concluded only two weeks before. He also suspended some $300 million in potential benefits to Israel through arms sales.[19]

According to the Reagan administration, the extension of Israeli law to the Golan constituted a violation of United Nations Security Council Resolution 242 – the resolution intended to serve as the basis of a permanent Arab–Israeli peace accord.

Some verbal fireworks notwithstanding, this American–Israeli disagreement soon turned into a tempest in a teapot. Despite the Reagan administration's sanctions, the Begin government held firm to its position on the Golan. A defiant Begin even unleashed a broadside against the administration's stance.

> What kind of expression is this – "punishing Israel"? Are we a vassal state of yours? Are we a banana republic? Are we youths of fourteen who, if they don't behave properly, are slapped across the fingers? . . . You will not frighten us with "punishments." He who threatens us will find us deaf to his threats. We are only prepared to listen to rational arguments. . . . You have no right to punish Israel – and I protest at the very use of this term.[20]

For its part, though Begin's remarks caused considerable anger within administration circles, the United States did not press Israel on the matter of the Golan law for very long. Strategic cooperation and arms transfers resumed after a brief interlude.

Just as they had in the Carter years, arms transfers to Arab states, specifically Saudi Arabia, troubled the American–Israeli relationship during the Reagan administration. But, unlike the Carter administration, which used arms transfers to the Jewish state in an effort to subdue the Begin government's opposition, the Reagan administration chose a much less confrontational alternative.

With respect to ancillary equipment for Saudi Arabian F-15s, the administration quietly negotiated a compromise with Israel. In return for the Begin government's acquiescence to the sale of additional items for these aircraft, the United States consented not to supply multiple-ejection bomb racks, the piece of equipment that represented the greatest threat to the Jewish state's security.

The same tactic, however, did not bear fruit with regard to the sale of Airborne Warning and Control System (AWACS) aircraft, sophisticated aerial platforms that could monitor military movements over land, sea, and air for hundreds of miles in all directions.[21] From the moment the Reagan administration expressed its intent to sell these aircraft to Saudi Arabia to the moment that it actually concluded the sale, the Begin government remained unalterably and publicly opposed to the deal, citing it as a threat to the Jewish state's security. An Israeli cabinet statement one day after the United States announced the consummation of the sale summed up the Begin government's long-standing position.

> The Government of Israel expresses its regret over the decision of the American Senate to approve the proposal of the [Reagan] administration on the two-fold arms deal between the United States and Saudi Arabia, which is in a state of war with Israel, rejects the Camp David Accords, and finances terror in our region. . . . The Government reiterates that a new and serious danger now faces Israel. . . .[22]

Despite the Begin government's vocal protests against the AWACS deal, the Reagan administration never attempted to use arms transfers to Israel to provide itself with leverage. During months of debate, it did not raise the prospect of an interruption in American weapons supplies to the Jewish state if the Begin government continued to object to the sale. The administration did hint that strategic cooperation with Israel might be dependent on a positive outcome of the AWACS debate; however, even on this issue, the United States chose not to confront the Jewish state too vigorously.

Israeli military operations in Iraq and Lebanon elicited a much stronger reaction from the Reagan administration. In late spring 1981, IAF F-16s bombed Iraq's Osirak nuclear reactor, completely destroying the facility. The Begin government claimed, correctly as evidence would later show, that Iraq sought to use this reactor to construct a nuclear arsenal. Shortly thereafter, Israel bombed PLO headquarters in Beirut in response to Palestinian attacks against the Jewish state.

The Reagan administration swiftly condemned both Israeli operations, moving beyond mere rhetoric.[23] The United States temporarily halted aircraft deliveries to the Jewish state pending an investigation into whether Israel had violated American end-use restrictions in bombing Osirak. The Reagan administration advanced the same pretext to continue the arms suspension after the Jewish state knocked out the PLO's headquarters.

The Begin government, not surprisingly, reacted heatedly to the Reagan administration's decisions, terming them absolutely unjustifiable. Nevertheless, Israel did agree to a cease-fire with the PLO in summer 1981, perhaps under the impression that it had to pay this price in order to get the United States to lift the embargo on aircraft shipments. Still, the influence on Israeli decision making proved short-lived, as the Jewish state invaded Lebanon in spring 1982 following PLO violations of the cease-fire accord.

Analysis of American and Israeli Conduct

The Nixon administration considered the Soviet Union to be an expansionist power that sought to spread its influence around the globe at Western expense. The Soviet Union's efforts to increase its presence in the Middle East caused particular alarm to the United States, because of the necessity of ensuring unhindered Western access to the region's oil reserves. Consequently, the administration's major foreign policy objective in the area revolved around the "containment" of the Soviet Union.

Israel constituted a strategic asset in this regard, according to the Nixon administration, especially after the role it had played in helping Jordan's pro-Western regime to survive that state's civil war. A strong Jewish state, the United States reasoned, served as a bulwark against Soviet and radical Arab troublemaking. Furthermore, by making Israel itself more secure, arms shipments discouraged the Jewish state from behaving in ways that could harm American interests in the region. Hence, the Nixon administration's decision to arm Israel from fall 1970 to fall 1973 without demanding meaningful Israeli concessions in return.

At least until the Soviet invasion of Afghanistan, the Carter administration displayed less concern with Soviet expansionism than any other Cold War American government. Rather than focus on the Soviet threat, this administration tended to operate on the premise that regional problems around the world largely lay outside the sphere of the East–West conflict. Indeed, the Carter administration even appeared to believe at times that the Soviet Union could play a beneficial role in solving these problems. It would not have invited the Soviet Union to cohost a renewed Geneva Peace Conference to settle the Arab–Israeli conflict if it thought otherwise.

The Reagan administration's attitude toward Soviet expansionism, to the contrary, paralleled the Nixonian view. To its way of thinking, the fingerprints of the Soviet Union were to be found all over every regional conflict. The Soviets, according to the Reagan administration, sought to stir up trouble around the globe both to advance their own national interests and to harm those of Western states. The primary purpose of American foreign policy, therefore, must be to confront and, wherever possible, roll back Soviet expansionism. This underlying logic accounts for the Reagan administration's emphasis on forming an anti-Soviet strategic consensus among Western-oriented states in the Middle East.

Paradoxically, despite their radically opposed *weltanschauungs*, the Carter and Reagan administrations conduct toward Israel evidenced certain striking similarities. The Carter administration generally viewed the Jewish state as a strategic nuisance. It laid much of the blame for the ongoing Arab–Israeli conflict at the feet of Israeli "intransigence," and it accused the Jewish state of heightening regional tensions through counterinsurgency operations in Lebanon. To move toward a solution of the Arab–Israeli conflict and to reduce Middle Eastern tensions, according to this perspective, the Begin government's behavior had to be actively channeled down the appropriate road. Arms transfers could serve as one mechanism to influence Israeli conduct. The administration, therefore, employed arms transfers alternately to punish or reward the Jewish state for its behavior.

The Reagan administration viewed Israel, the strongest military power in the Middle East, as a valuable strategic asset in the quest to block Soviet expansionism in the area. But it also viewed Western-oriented Arab states, such as Saudi Arabia and Egypt, as important cogs in its cherished anti-

Soviet regional alignment. Thus, on the one hand, the Reagan administration used arms transfers to strengthen the Jewish state in order to enhance Israel's capability to act as an anti-Soviet bulwark. On the other hand, though, the administration also employed arms transfers to sanction Israel, less because it was truly disturbed by the Begin government's conduct *vis-à-vis* the Golan, Iraq, and Lebanon and more because it did not want to alienate its Arab clients. Like its predecessor, then, the Reagan administration ultimately adopted a carrot-and-stick approach to arms transfers to the Jewish state, but to serve a very different foreign policy agenda.

Historically speaking, Israel proved far less responsive to American arms pressure during the (peacetime) Nixon, Carter, and Reagan years than it had during the Johnson, (wartime) Nixon, and Ford years. Two basic trends account for this fact – one external to the Jewish state and one internal to it. Externally, the (peacetime) Nixon, Carter, and Reagan administrations, whatever disagreements they may have had with the Meir and Begin governments, typically did not wield the arms instrument with the same overall zeal as the Johnson, (wartime) Nixon, and Ford administrations, because, in the absence of full-scale war in the Middle East, the stakes for the United States were typically not as high. The potential for dramatic foreign policy reversals or gains in peacetime were simply not as great as they were in wartime, so the Carter and, especially, the (peacetime) Nixon and Reagan administrations felt somewhat less urgency in making Israel toe the American line than their counterparts.

Internally, in the absence of full-scale war, the Jewish state considered itself to be reasonably secure. The Meir and Begin governments thought, therefore, that they did not need to make major diplomatic concessions to the (peacetime) Nixon, Carter, and Reagan administrations in return for American arms transfers. So long as it did not face a massive Arab attack, Israel had sufficient stockpiles of weapons to weather any possible storms. Consequently, the Meir and Begin governments displayed some willingness to compromise over what were for the Jewish state more marginal concerns – procedural issues with respect to an interim agreement with Egypt over the Suez Canal and with respect to the proposed Geneva Peace Conference, as well as the scope and duration of Israeli counterinsurgency operations in Lebanon, for example – in exchange for arms. But they would not budge on more significant concerns – American weapons sales to Arab states and Israeli policy in the administered territories, for example – in return for those arms.

Conclusion

The Costs of an Alliance and the Benefits of a Patron–Client Relationship

Beginning in the 1950s, prominent officials in both the United States and Israel have occasionally floated the idea of an American–Israeli alliance.[1] But the vast majority of both American and Israeli statesmen over the decades have never given serious thought to this sort of arrangement.[2] Rather, they have seemingly been content to live with a patron–client relationship. The question, therefore, is: why have both the United States and Israel favored an informal over a formal relationship?

The American Perspective

The United States has always had national interests on both sides of the Arab–Israeli conflict. With respect to the Arab world, the primary American interest has been to assure Western access to oil reserves. Furthermore, before the end of the Cold War and the dissolution of the Soviet Union, the United States also believed that to limit the extent of Soviet penetration of the Arab world constituted an important interest. With regard to Israel, the basic American interest has been to ensure the survival of the Jewish state.

To protect its national interests in both the Arab world and Israel, the United States has steadfastly sought to avoid any action that could lead to the perception that it has chosen in an official sense to back either side over the other one. An alliance with Israel, of course, would immediately be interpreted in the Arab world as a declaration that the United States had decided to stand squarely behind the Jewish state in the Arab–Israeli conflict. Consequently, successive administrations have been quite unenthusiastic about the prospect of an American–Israeli alliance.

Moreover, the United States has been concerned that an American–Israeli alliance could lead to its entrapment.[3] To put it another way, the United States has feared that it could be dragged directly into an

Arab–Israeli war as a result of binding obligations arising out of an alliance with the Jewish state. Because the United States has reasoned that involvement in such a war would be disastrous for American interests in the Middle East – and perhaps around the world – it has been very wary of entering into an alliance with Israel.

Measured against these potential costs, the potential benefits of an American–Israeli alliance have appeared to be rather modest. The United States, to be sure, would acquire enhanced influence over Israeli foreign policy; however, through the manipulation of arms transfers, as well as the application of other forms of pressure, it has exercised significant control over the Jewish state's policy in the past, especially during Middle Eastern wars. Time and again, the United States has proved able to wring concessions out of Israel in order to advance American national interests.

Furthermore, the United States has reaped abundant benefits in the absence of an alliance. First, over the course of its existence, Israel has provided to the United States a regular stream of high quality intelligence data about countries in the Middle East, Africa, and the former Soviet bloc.[4] Nikita Khrushchev's secret speech to fellow Communist Party leaders denouncing Stalinist era crimes is but one of the many valuable items of information passed on to the United States by Israel's intelligence services. Likewise, the Jewish state has routinely shared technical and tactical information gained from its military operations with the United States.[5] This information has been effectively utilized by American armed forces in a variety of places, including Vietnam, Grenada, and Panama. Recently, in Iraq, the United States has employed counterinsurgency tactics derived from Israeli operations in Judea, Samaria, and Gaza.

The Jewish state has also pledged to host American armed forces on its soil in a future emergency, and it has even modified certain of its facilities to accommodate the needs of these forces. Finally, Israel has occasionally acted either as a substitute for American power in the Middle East – for example, during the 1970 Jordanian civil war – or as an American "beard" that has supported states or groups – for example, in Latin America and the Middle East – with which the United States itself has preferred not to have a direct relationship. The fact that the United States has enjoyed these benefits in the absence of an alliance has lowered still further its incentive to seek such an arrangement with the Jewish state.

The Israeli Perspective

The Jewish state has been concerned that an American–Israeli alliance could harm its national security by reducing its defensive capabilities and leaving it vulnerable to abandonment.[6] Israel has worried that, as the price of an alliance, the United States could require it to divest itself of vital security assets. First, the United States could demand that the Jewish state give up its insistence on defensible borders and return to the vulnerable pre-1967 Six-Day War frontiers. Second, the United States could insist that the Jewish state relinquish its nuclear option. And, third, the United States could stipulate that Israel make do with smaller quantities of American weapons (and other forms of material assistance). The extension of an American shield over Israel through the vehicle of an alliance, the United States might argue, would serve as a reliable substitute for these assets.

After the Jewish state had been stripped of territory, nuclear weapons, and American arms, Israel has feared, the United States could then refuse to honor its alliance commitment. The events of 1967 – when the United States refused to guarantee the Jewish state's right of free passage in international waters, even though it had earlier promised to do so – are never far from Israeli minds in this context. Israel has reasoned, in short, that an American–Israeli alliance could leave it with far fewer local capabilities with which to protect itself in an hour of abandonment than it would otherwise have had by refusing to enter into such an arrangement in the first place.

The Jewish state has also been concerned that an alliance with the United States could further reduce its freedom of action.[7] Every Israeli foreign policy decision that could conceivably affect American national interests would have to be cleared with the United States. If the Jewish state wanted to strike a terrorist training center in a neighboring state, for example, it would have to seek American permission to launch an attack. As the stronger member of an asymmetric alliance, the United States would exercise veto power over any Israeli decision that it deemed inimical to its interests.

Finally, like the United States, Israel has had to consider how an American–Israeli alliance might be viewed by third parties. Many countries around the globe, including major powers such as the People's Republic of China and Russia, have long been suspicious of the United States. An American–Israeli alliance could hurt the Jewish state's relationships with these countries by reinforcing the notion that it is a willing servant of American national interests around the world.

Weighed against the potential costs of an alliance, the potential benefits have appeared to be very limited in Israeli eyes. An alliance would certainly symbolize America's commitment to the Jewish state; however, because Israel's most dangerous foes are already convinced that the United States stands solidly behind the Jewish state, such a symbol would seem to be of no great deterrent value.

Moreover, Israel has derived considerable advantages from the patron–client relationship. The United States has furnished the Jewish state with copious quantities of arms, in part because America has been unwilling to provide a formal security guarantee to Israel. For much the same reason, the United States has also largely looked the other way from the 1960s onward as the Jewish state developed a potent nuclear arsenal, despite the fact that nonproliferation has long been a major tenet of American foreign policy. Likewise, America's position that a comprehensive Arab–Israeli peace should leave Israel with defensible borders, whatever differences the Jewish state and the United States may have over the precise location of those frontiers, reflects the understanding that the latter cannot strip the former of its ability to defend itself in the absence of an American–Israeli alliance. In light of these benefits, Israel simply has not been in any hurry to formalize its relationship with the United States.

Into the Future

To forecast the future on the basis of the past, of course, is always a risky enterprise. Nevertheless, the United States and Israel are unlikely to move from a patron–client relationship to an alliance any time soon. Not only will they continue to perceive the present relationship as accommodating their national interests quite satisfactorily, but they will also continue to operate under the assumption that the potential costs of an alliance outweigh the potential benefits of one. The security-for-autonomy bargain at the heart of the American–Israeli patron–client relationship, therefore, is likely to last into the indefinite future.

Notes

Introduction: The American–Israeli Relationship in Historical Perspective

1. Abba Eban, *The New Diplomacy: International Affairs in the Modern Age* (New York: Random House, 1983), p. 215.
2. Many observers who are familiar with the history of the relationship employ this term to describe it. On the basis of his very broad definition of an alliance as a "formal or informal arrangement for security cooperation between [or among] two or more sovereign states," Stephen Walt applies this term to the American–Israeli relationship. See Stephen M. Walt, *The Origins of Alliances* (Ithaca, NY: Cornell University Press, 1987), p. 12. In a less theoretical context, Warren Bass also uses this term to describe the American–Israeli relationship. See Warren Bass, *Support Any Friend: Kennedy's Middle East and the Making of the U.S.–Israel Alliance* (New York: Oxford University Press, 2003).
3. Charles W. Kegley, Jr. and Gregory A. Raymond, *When Trust Breaks Down: Alliance Norms and World Politics* (Columbia, SC: University of South Carolina Press, 1990), p. 52. Ole Holsti, P. Terrance Hopmann, and John D. Sullivan define an alliance in the same vein. In their view, "an alliance is a formal agreement between [or among] two or more nations to collaborate on national security issues." See Ole Holsti, P. Terrance Hopmann, and John D. Sullivan, *Unity and Disintegration in International Alliances* (New York: John Wiley & Sons, 1973), p. 4.
4. A patron–client relationship, however, does bear more than a passing similarity to an "asymmetric alliance." Members of such an alliance incur different costs and derive different benefits from this arrangement. The smaller member receives security benefits from the larger member at a certain cost to the smaller member's autonomy, while the larger member receives autonomy benefits from the smaller member at a certain cost to the larger member's security. Put differently, the smaller member of an asymmetric alliance cedes some of its freedom to pursue its foreign policy agenda in order to strengthen its military capabilities, while the larger member cedes some of its military capabilities in order to increase its freedom to pursue its foreign policy agenda. See James D. Morrow, "Alliances and Asymmetry: An Alternative to the Capability Aggregation Model of Alliances," *American Journal of Political Science*, Vol. 35, No. 4 (November 1991), pp. 904–933.
5. For the particular reasons why both the United States and Israel have preferred a patron–client relationship to an alliance see David Rodman, "Patron–Client

Dynamics: Mapping the American–Israeli Relationship," *Israel Affairs*, Vol. 4, No. 2 (Winter 1997), pp. 38–39. Broadly speaking, the United States has not wanted to tie itself too closely to Israel, while Israel has not wanted to make itself too submissive to the United States. For a more detailed explanation see the Conclusion.

6 For accounts of the American–Israeli relationship in the Truman years see Michael J. Cohen, *Truman and Israel* (Berkeley: University of California Press, 1990); Peter L. Hahn, *Caught in the Middle: U.S. Policy Toward the Arab–Israeli Conflict, 1945–1961* (Chapel Hill, NC: The University of North Carolina Press, 2004); and Steven L. Spiegel, *The Other Arab–Israeli Conflict: Making America's Middle East Policy, from Truman to Reagan* (Chicago: The University of Chicago Press, 1985).

7 For the amount of economic assistance furnished to Israel during the Truman administration see A. F. K. Organski, *The $36 Billion Bargain: Strategy and Politics in U.S. Assistance to Israel* (New York: Columbia University Press, 1990), p. 142.

8 American arms transfers to Israel will be addressed in depth in subsequent chapters.

9 Even though the United States gave strong diplomatic backing to the pact, it refused to join, perhaps because it preferred not to make formal military commitments in the Middle East that might compromise its ability to defend Western Europe against a potential Soviet invasion. On American cautiousness about allocating military resources to the Middle East in the early Cold War years see Michael J. Cohen, *Fighting World War III from the Middle East: Allied Contingency Plans, 1945–1954* (London: Frank Cass, 1997).

10 Israeli counterinsurgency operations inside the Gaza Strip in response to Palestinian terrorist attacks against Israel, especially one devastating raid in early 1955 that killed about 40 Egyptian soldiers, also contributed to Egypt's decision to seek assistance from the Soviet Union. See, for example, Motti Golani, *Israel in Search of a War: The Sinai Campaign, 1955–1956* (Brighton & Portland: Sussex Academic Press, 1998).

11 For accounts of the American–Israeli relationship during the first Eisenhower administration see Isaac Alteras, *Eisenhower and Israel: U.S.–Israeli Relations, 1953–1960* (Gainesville, FL: University Press of Florida, 1993); Hahn, *Caught in the Middle*; and Spiegel, *The Other Arab–Israeli Conflict*.

12 For the Eisenhower administration's use of coercive diplomacy to get Israel to end its water construction project and to withdraw from the Sinai see Abraham Ben-Zvi, *The United States and Israel* (New York: Columbia University Press, 1993).

13 For the amount of economic assistance see Organski, *The $36 Billion Bargain*, p. 142.

14 On the Alpha Plan see Orna Almog, *Britain, Israel, and the United States, 1955–1958: Beyond Suez* (London: Frank Cass, 2003).

15 For accounts of the American–Israeli relationship during the second Eisenhower administration see Alteras, *Eisenhower and Israel*; Abraham Ben-Zvi, *Decade of Transition: Eisenhower, Kennedy, and the Origins of the*

American–Israeli Alliance (New York: Columbia University Press, 1998); Hahn, *Caught in the Middle*; and Spiegel, *The Other Arab–Israeli Conflict.*

16 This point is made in Michael Mandelbaum, *The Fate of Nations: The Search for National Security in the Nineteenth and Twentieth Centuries* (New York: Cambridge University Press, 1988), p. 284.

17 On the Lebanese and Jordanian crises see Almog, *Britain, Israel, and the United States* and Ben-Zvi, *Decade of Transition.*

18 For accounts of the American–Israeli relationship during the Kennedy administration see Bass, *Support Any Friend*; Ben-Zvi, *Decade of Transition*; Abraham Ben-Zvi, *John F. Kennedy and the Politics of Arms Sales to Israel* (London: Frank Cass, 2002); Mordechai Gazit, *President Kennedy's Policy Toward the Arab States and Israel: Analysis and Documents* (Tel Aviv: Shiloah Center for Middle Eastern and African Studies, 1983); David Schoenbaum, *The United States and the State of Israel* (New York: Oxford University Press, 1993); and Spiegel, *The Other Arab–Israeli Conflict.*

19 The Egyptian destabilization campaign, as well as the American reaction to it, is chronicled in Zaki Shalom, *The Superpowers, Israel and the Future of Jordan, 1960–1963: The Perils of the Pro-Nasser Policy* (Brighton & Portland: Sussex Academic Press, 1999).

20 The most thorough examination of this sale is to be found in Ben-Zvi, *John F. Kennedy and the Politics of Arms Sales to Israel.* See also Gazit, *President Kennedy's Policy Toward the Arab States and Israel.*

21 For accounts of the American–Israeli relationship during the Johnson administration see Schoenbaum, *The United States and the State of Israel* and Spiegel, *The Other Arab–Israeli Conflict.*

1 The Israeli Quest for Arms: Western Europe and the United States

1 An overview of the Yishuv's arms policy can be found in Stewart Reiser, *The Israeli Arms Industry: Foreign Policy, Arms Transfers, and Military Doctrine of a Small State* (New York: Holmes & Meier, 1989), pp. 1–16.

2 On the embargo see Amitzur Ilan, *The Origin of the Arab–Israeli Arms Race: Arms, Embargo, Military Power and Decision in the 1948 Palestine War* (New York: New York University Press, 1996) and Shlomo Slonim, "The 1948 American Embargo on Arms to Palestine," *Political Science Quarterly*, Vol. 94, No. 3 (Fall 1979), pp. 495–514.

3 This unintended result is a central theme of Ilan, *The Origin of the Arab–Israeli Arms Race.*

4 The Israeli endeavor to circumvent the arms (and manpower) embargo is well known. See David J. Bercuson, *The Secret Army* (New York: Stein and Day, 1984); Joseph A. Heckelman, *American Volunteers and Israel's War of Independence* (New York: KTAV Publishing House, 1974); Leonard Slater, *The Pledge* (New York: Simon & Schuster, 1970); and Jeffrey Weiss and Craig Weiss, *I Am My Brother's Keeper: American Volunteers in Israel's War for Independence* (Atglen, PA: Schiffer Military History, 1998).

5 Late in the war, in order to coerce Israel into withdrawing from the northern Sinai, which the Jewish state had seized during its final offensive operations

against the Egyptian army in the Negev, Great Britain even threatened to join the fighting on the Arab side by invoking the terms of an Anglo–Egyptian defense treaty.

6 On the Israeli–Soviet–Czech arms connection see Uri Bialer, *Between East and West: Israel's Foreign Policy Orientation* (Cambridge: Cambridge University Press, 1990), pp. 173–180. The Soviets, who approved of arms sales to both Israel and the Arab world during the war in an attempt to weaken Western influence in the Middle East and to extend their own in the area, did not want to be seen as publicly flouting the embargo; hence, they had Czechoslovakia sell the arms.

7 The text of the Tripartite Declaration is reprinted as part of The Avalon Project at the Yale Law School <www.yale.edu/lawweb/avalon>.

8 On Israel's brief dalliance with nonalignment see Bialer, *Between East and West.*

9 Israel's requests for arms and search for a security guarantee in these years are chronicled in Zach Levey, *Israel and the Western Powers, 1952–1960* (Chapel Hill, NC: The University of North Carolina Press, 1997), pp. 7–34. Also see Isaac Alteras, *Eisenhower and Israel: U.S.–Israeli Relations, 1953–1960* (Gainesville, FL: University Press of Florida, 1993).

10 On Israel's defense relationships with Great Britain and France in the early 1950s see Orna Almog, *Britain, Israel, and the United States, 1955–1958: Beyond Suez* (London: Frank Cass, 2003); Mordechai Bar-On, *The Gates of Gaza: Israel's Road to Suez and Back, 1955–1957* (New York: St. Martin's Press, 1994); Sylvia K. Crosbie, *A Tacit Alliance: France and Israel from Suez to the Six-Day War* (Princeton: Princeton University Press, 1974); and Levey, *Israel and the Western Powers*, pp. 35–74.

11 For Israel's rejection of Great Britain's offer see Michael J. Cohen, *Fighting World War III from the Middle East: Allied Contingency Plans, 1945–1954* (London: Frank Cass, 1997) and Levey, *Israel and the Western Powers*, pp. 35–54.

12 Israel's requests for arms and search for a security guarantee in these years are chronicled in Abraham Ben-Zvi, *Decade of Transition: Eisenhower, Kennedy, and the Origins of the American–Israeli Alliance* (New York: Columbia University Press, 1998) and Zach Levey, *Israel and the Western Powers*, pp. 80–100. Also see Isaac Alteras, *Eisenhower and Israel.*

13 For this sale see Ben-Zvi, *Decade of Transition*, p. 83 and Levey, *Israel and the Western Powers*, p. 98.

14 On Israel's defense relationships with Great Britain, France, and the Federal Republic of Germany in the post-Sinai Campaign years see Almog, *Britain, Israel, and the United States*; Crosbie, *A Tacit Alliance*; Lily Gardner Feldman, *The Special Relationship Between West Germany and Israel* (Boston: Allen & Unwin, 1984); and Levey, *Israel and the Western Powers*, pp. 101–131.

15 For French participation in Israel's nuclear weapons and SSM programs see Avner Cohen, *Israel and the Bomb* (New York: Columbia University Press, 1998).

16 For a list of specific weapons furnished to Israel by the Federal Republic of

Germany see telegram, United States Embassy in the Federal Republic to Department of State (DoS), 15 February 1965, "Israel, Tanks Vol. 2," Country File, National Security File, Box 145, LBJ Library.

17 Israel's requests for arms and a security guarantee during the Kennedy administration are traced in Abraham Ben-Zvi, *John F. Kennedy and the Politics of Arms Sales to Israel* (London: Frank Cass, 2002). See also Mordechai Gazit, *President Kennedy's Policy Toward the Arab States and Israel: Analysis and Documents* (Tel Aviv: Shiloah Center for Middle Eastern and African Studies, 1983).

18 On the connection between Israel's progress in the nuclear field and the transfer of American arms during the 1960s see Cohen, *Israel and the Bomb* and Zaki Shalom, *Israel's Nuclear Option: Behind the Scenes Diplomacy Between Dimona and Washington* (Brighton & Portland: Sussex Academic Press, 2005).

19 For an account of the Johnson administration's arms deals with Israel see Abraham Ben-Zvi, *Lyndon B. Johnson and the Politics of Arms Sales to Israel: In the Shadow of the Hawk* (London: Frank Cass, 2004).

2 Armored Breakthrough: The 1965 Sale of M-48 Patton Tanks to Israel

1 The terms of the deal are outlined in a 29 July 1965 memorandum (230) from National Security Council staff member Robert Komer to President Lyndon Johnson. This document is contained in *Foreign Relations of the United States, 1964–1968*, Volume XVIII, Arab–Israeli Dispute, 1964–67. All of the documents cited in this chapter are contained in this volume unless noted to the contrary. (Each document's reference number in Volume XVIII appears in parentheses after its date of circulation.)

2 Two years after the sale, during the 1967 Six-Day War, M-48s would participate in the IDF's strike against Egypt in the Sinai.

3 On the Skyhawk sale see chapter 3.

4 On the Phantom sale see chapter 4.

5 Avner Cohen, *Israel and the Bomb* (New York: Columbia University Press, 1998), p. 198.

6 The Israeli position is summarized in a 3 January 1964 memorandum of conversation (3) circulated by Assistant Secretary of State for Near Eastern and South Asian Affairs Phillips Talbot.

7 The assessments prepared by the JCS are dated 18 January 1964 (10), 12 March 1964 (28), and 6 May 1965 (211).

8 Bundy's opinion appears in Komer's 10 January 1964 memorandum for the record (7).

9 For the Secretary of Defense's perspective see Solbert's 15 February 1964 memorandum (13).

10 For the Secretary of State's perspective see his 16 January 1964 and 25 February 1964 memoranda (9 and 18) to Johnson.

11 Johnson's memorandum to Feldman is dated 15 May 1964 (55).

12 See Feldman's 14 March 1964 memorandum (29) to Johnson. For the contrary view see Bundy's 8 March 1964 memorandum (27) to Johnson.

McNamara and Rusk, it should be noted, displayed a greater willingness to involve the United States in a tank deal sooner rather than later in comparison to the NSC, DoS, DoD, and CIA experts working on this issue.

13 On these programs see Cohen, *Israel and the Bomb*. The SSM program would eventually result in the Jericho family of missiles, while the nuclear program saw at least two bombs readied for use on the eve of the Six-Day War.

14 These lines are taken from Johnson's 20 February 1964 letter (14) to Eshkol. The link between Israeli SSMs/nuclear weapons and American tanks is rendered unambiguously in many documents. See, for example, Rusk's 16 January 1964 memorandum (9) to Johnson and the joint DoS–DoD 25 April 1964 memorandum (47).

15 The Eshkol government's position that the acquisition of tanks should not be connected to Israel's SSM and nuclear research programs is reflected in Komer's 5 March 1964 memorandum for the record (26).

16 This theme, too, appears in many documents. See Deputy Assistant Secretary of State for Near Eastern and South Asian Affairs John Jernegan's 28 February 1964 memorandum (21) to Deputy Under Secretary of State for Political Affairs U. Alexis Johnson and the Special National Intelligence Estimate of 15 April 1964 (42). Also see the telegrams from United States embassies throughout the Arab world, which can be found in "Israel, Tanks Vol. 1," Country File, National Security File, Box 145, LBJ Library.

17 This attitude is forthrightly expressed, for example, in Rusk's 16 January 1964 memorandum (9) to Johnson as well as in his 20 August 1964 telegram (92) to the United States Embassy in Jordan.

18 For this recommendation see the late April 1964 memoranda (47 and 49) outlining joint DoS–DoD thinking.

19 The administration's support for the plan is evident in memoranda among senior officials dated 15 May 1964 (55), 16 May 1964 (57), and 28 May 1964 (63).

20 The Johnson–Eshkol exchange of views is summarized in a 1 June 1964 memorandum of conversation (65).

21 *Ibid.*

22 Egypt had shown no genuine willingness to halt its SSM and chemical warfare programs in its contacts with the Johnson administration.

23 The Israeli position on acquiring tanks in Western Europe can be gleaned from memoranda dated 17 May 1964 (58) and 2 June 1964 (67).

24 For West Germany's reluctance see Acting Assistant Secretary of Defense for International Security Affairs John McNaughton's 16 May 1964 memorandum (56) to McNamara. Also see Steven L. Spiegel, *The Other Arab–Israeli Conflict: Making America's Middle East Policy, from Truman to Reagan* (Chicago: The University of Chicago Press, 1985), p. 132. For arms links between the Federal Republic of Germany and Israel from the late 1950s to the mid-1960s see Lily Gardner Feldman, *The Special Relationship Between West Germany and Israel* (Boston: Allen & Unwin, 1984), pp. 122–136. For a list of arms transferred to Israel by the Federal Republic see telegram, United

States Embassy in the Federal Republic to DoS, 15 February 1965, "Israel, Tanks Vol. 2," Country File, National Security File, Box 145, LBJ Library.

25 On West German fears see Feldman, *The Special Relationship Between West Germany and Israel,* p. 161.

26 The terms of the American–Israeli–West German tank deal can be found in memoranda dated 23 September 1964 (95) and 19 October 1964 (99).

27 See Rusk's telegrams to the United States Embassy in Egypt dated 29 February 1964 (22) and 3 May 1964 (50).

28 See, for instance, the telegram from the United States Embassy in Egypt to the DoS dated 4 March 1964 (24).

29 Jordan's appetite for American arms is foreshadowed in a 15 April 1964 memorandum of conversation (40).

30 Feldman, *The Special Relationship Between West Germany and Israel,* p. 182.

31 For this Israeli demand see the DoS's 13 February 1965 telegram (148) to the United States Embassy in Israel and Komer's 16 February 1965 memorandum (152) to Johnson.

32 Historical background information on the Jordan River water problem can be found in Michael Brecher, *Decisions in Israel's Foreign Policy* (New Haven: Yale University Press, 1975), pp. 173–213 and Avner Yaniv, *Deterrence Without the Bomb: The Politics of Israeli Strategy* (Lexington, MA: Lexington Books, 1987), pp. 104–107.

33 Eshkol's Knesset speech is available at the Israel Ministry of Foreign Affairs web site <www.mfa.gov.il>.

34 This communiqué is also archived at the Ministry of Foreign Affairs web site.

35 The American objection is reflected in numerous memoranda and telegrams, including those dated 13 February 1965 (149), 16 February 1965 (152), 21 February 1965 (157), 26 February 1965 (160), 28 February 1965 (168), 1 March 1965 (169), and 3 March 1965 (175).

36 For this border skirmishing see Yaniv, *Deterrence Without the Bomb*, pp. 106–107.

37 See the memoranda to Rusk dated 22 July 1964 (78) and 24 July 1964 (79).

38 On Jordan's position see the memorandum to Rusk dated 22 July 1964 (78).

39 For the administration's concern see the Special National Intelligence Estimate of 13 August 1964 (91), the memorandum of conversation dated 14 January 1965 (121), and Komer's memorandum to Johnson dated 21 January 1965 (124).

40 The thinking of American officials is outlined in Rusk's 1 February 1965 memorandum (129) to Johnson, the NSC's 1 February 1965 meeting summary (130), and the DoS's 1 February 1965 telegram (131) to the United States Embassy in Israel.

41 The course of American–Jordanian negotiations over the arms sale can be traced in a series of memoranda and telegrams dated 7 February 1965 (140 and 141), 9 February 1965 (143 and 145), 13 February 1965 (149), 19 February 1965 (154 and 155), and 12 March 1965 (188).

42 See memorandum, Komer to Johnson, 7 February 1965, "Israel, Tanks Vol.

2," Country File, National Security File, Box 145, LBJ Library.

43 The Eshkol government's response is cited in two DoS telegrams dated 6 February 1965 (137) and 13 February 1965 (147).

44 See the DoS telegram of 13 February 1965 (148) and Komer's 16 February 1965 memorandum (152) to Johnson.

45 Under Secretary of State George Ball and Komer were two key officials who held this opinion. See Ball's telegram to the United States Embassy in Jordan dated 8 February 1965 (142), as well as Komer's memoranda to Johnson dated 6 February 1965 (138), 9 February 1965 (143), and 16 February 1965 (152).

46 These guidelines are contained in Johnson's 21 February 1965 memorandum (157) to Harriman and Komer. Also see Rusk's 26 February 1965 telegram (160) to the United States Embassy in Israel.

47 The Eshkol government's reaction to the American proposal, as well as its counterproposal, appears in a pair of telegrams sent to the DoS by American Ambassador to Israel Walworth Barbour. These telegrams summarize Harriman and Komer's talks with Israeli officials. The telegrams are dated 26 February 1965 (161) and 27 February 1965 (163).

48 The Harrison–Komer recommendation is contained in a pair of telegrams sent to the DoS by Barbour. These telegrams are dated 28 February 1965 (165) and 2 March 1965 (173). See also Komer's messages to Bundy dated 1 March 1965 (172) and 6 March 1965 (180).

49 See Rusk's telegrams to the United States Embassy in Israel dated 3 March 1965 (175), 7 March 1965 (181), and 8 March 1965 (182).

50 The complete text of this MoU appears in Barbour's 11 March 1965 telegram (185) to the DoS.

51 See Rusk's 10 March 1965 telegram (186) to the United States Embassy in Jordan and his 18 March 1965 telegram (193) to the United States Embassy in Egypt.

52 A hint of the ongoing wrangling can be detected in Komer's 20 May 1965 memorandum for the record (216).

3 One Step Forward and One Step Backward: The 1966 Sale of A-4 Skyhawk Aircraft to Israel and the Post-1967 Six-Day War Arms Embargo

1 The terms of the agreement are outlined in Komer's 22 February 1966 memorandum (273) to Johnson. This document is contained in *Foreign Relations of the United States, 1964–1968*, Volume XVIII, Arab–Israeli Dispute, 1964–67. All of the documents cited in notes 1–23 are contained in this volume unless noted to the contrary. (Each document's reference number in Volume XVIII appears in parentheses after its date of circulation.)

2 For Israel's request that state-of-the-art ordnance and electronic systems be included in the aircraft package see Special Assistant to the President Walt Rostow's 3 May 1966 memorandum for the record (288). For the Johnson administration's apparent unwillingness to furnish these items – at least for the time being – see the 22 August 1966 letter (316) from Assistant Secretary of

State for Near Eastern and South Asian Affairs Raymond Hare to Deputy Assistant Secretary of Defense for International Security Affairs Townsend Hoopes.

3 See the 17 May 1964 memorandum of conversation (58) about tanks.
4 See American Ambassador to Israel Walworth Barbour's telegrams from the United States Embassy in Israel to the DoS dated 27 February 1965 (163) and 28 February 1965 (168).
5 See Barbour's 11 March 1965 telegram (185), which spells out the terms of the American–Israeli MoU, to the DoS.
6 See Barbour's 27 February 1965 telegram (163) to the DoS.
7 The JCS view is revealed in a 6 May 1965 memorandum (211) to McNamara.
8 See Komer's 25 October 1965 memorandum (246) to Johnson.
9 The administration's position on an aircraft sale to Israel during these months is made abundantly transparent in a series of memoranda and telegrams dated 20 May 1965 (216), 5 June 1965 (222), 16 June 1965 (224), 29 July 1965 (230), 8 September 1965 (236), 30 September 1965 (238), 25 October 1965 (246), and 29 December 1965 (260).
10 See the various letters, memos, and telegrams circulated among American officials that describe the discussions between them and their British, French, and Israeli counterparts throughout autumn 1965 in "Israel Security, Arms/Aircraft Vols. 1 and 2 (1965)," Komer File, National Security File, Box 31, LBJ Library.
11 *Ibid.*
12 *Ibid.*
13 See Komer's 25 October 1965 memorandum (246) to Johnson.
14 See the various letters, memos, and telegrams circulated among American officials that describe the discussions between them and their British, French, and Israeli counterparts throughout autumn 1965 in "Israel Security, Arms/Aircraft Vols. 1 and 2 (1965)," Komer File, National Security File, Box 31, LBJ Library.
15 See Komer's 25 October 1965 memorandum (246) to Johnson.
16 See Komer's 12 January 1966 memorandum (261) to Johnson and his 21 January 1966 memorandum (263) to the Special Assistant to the President for National Security Affairs McGeorge Bundy.
17 See their joint January–February 1966 memorandum (262) to Johnson and Rusk's 24 January 1966 memorandum for the record (264).
18 See Komer's 8 February 1966 memorandum (267) to Johnson.
19 See the 9 February 1966 memorandum for the record (268).
20 The Johnson administration cited four reasons for refusing the Intruder to Israel: (1) its high cost; (2) its classified electronic systems; (3) its ability to deliver nuclear weapons; and (4) its production schedule for the navy, which would be interrupted by a sale to the IAF. See the 11 February 1966 memorandum for the record (270).
21 See the 12 February 1966 memorandum of conversation (271) and Komer's 22 February 1966 memorandum (273) to Johnson.
22 *Ibid.*

23 See the 31 March 1966 memorandum (283) from Assistant Secretary of Defense for International Affairs John McNaughton to McNamara.
24 See the 8 June 1967 memorandum (222) prepared by National Security Council Staff member Harold Saunders, *Foreign Relations of the United States, 1964–1968*, Volume XIX, Arab–Israeli Crisis and War, 1967. Also see the 9 June 1967 memorandum (235) from the President's Special Consultant McGeorge Bundy to Johnson. All of the documents cited in notes 24–42 are found in Volume XIX. (Each document's reference number in Volume XIX appears in parentheses after its date of circulation.)
25 On the UN registry idea see the 12 June 1967 meeting notes (269) of the Special Committee of the NSC and the 12 June 1967 memorandum for the record (270).
26 For the lack of Western European support for regional weapons limitations see the 15 June 1967 memorandum (293) for the Special Committee.
27 On the Soviet refusal see the 16 June 1967 memorandum of conversation (301), the 23 June 1967 memorandum of conversation (321), the 8 July 1967 memorandum of conversation (347), and the 27 July 1967 memorandum of conversation (392).
28 See the 29 September 1967 telegram (451) from the DoS to the United States Embassy in Israel. This recognition, it is worth noting, did not prevent the Johnson administration from revisiting the theme of arms control with the Soviets in the future, especially as the United States pondered the sale of the Phantom to Israel in 1968.
29 See the 16 June 1967 telegrams (266 and 286), as well as the 30 June 1967 telegram (333), from the DoS to the United States Embassy in Israel.
30 See the 15 July 1967 memorandum of conversation (369) and the 21 July 1967 telegram (381) from the DoS to the United States Embassy in Israel.
31 For Israeli requests and suspicions see the 7 August 1967 telegram (409) from the United States Embassy in Israel to the DoS, the 15 August 1967 memorandum for the record (418), the 16 August 1967 telegram (420) from the United States Embassy in Israel to the DoS, the 22 September 1967 memorandum (445) from the President's Special Assistant Walt Rostow to Johnson, the 26 September 1967 telegram (449) from the United States Mission to the UN to the DoS, the 4 October 1967 memorandum (457) from Saunders to Rostow, and the 7 October 1967 telegram (458) from the DoS to the United States Embassy in Israel.
32 On Israeli arms requests before the war see the 1 June 1967 telegram (101) from the DoS to the United States Embassy in Israel.
33 On the transfer of these gas masks see the 22 May 1967 memorandum (37) from the Deputy Assistant Secretary of Defense for International Security Affairs Townsend Hoopes to Secretary of Defense Robert McNamara.
34 See the 12 September 1967 memorandum (436) from Rostow to Johnson.
35 See the 16 June 1967 telegram (304) from the DoS to the United States Embassy in Jordan and the 28 June 1967 memorandum of conversation (331).
36 On the Jordanian requests see the 15 July 1967 memorandum of meeting

(368), the 24 October 1967 telegram (486) from the DoS to the United States Embassy in Jordan, and the 8 November 1967 telegram (508) from the DoS to the United States Mission to the UN.

37 See the 27 June 1967 memorandum (330) from Bundy to Johnson, the 6 July 1967 telegram (345) from Rostow to Johnson, and the 11 July 1967 memorandum (350) from Bundy to Johnson.

38 On these sales see the 27 June 1967 memorandum (330) from Bundy to Johnson, the 1 August 1967 memorandum for the record (403), and the 3 August 1967 memorandum (406) from Rostow to Johnson.

39 David Schoenbaum, *The United States and the State of Israel* (New York: Oxford University Press, 1993), p. 163 and Steven L. Spiegel, *The Other Arab–Israeli Conflict: Making America's Middle East Policy, from Truman to Reagan* (Chicago: The University of Chicago Press, 1985), p. 159.

40 For these views see the 10 August 1967 special national intelligence estimate (414) and the 25 August 1967 memorandum (427) from the JCS to McNamara.

41 See the 5 July 1967 memorandum of conversation (343), the 30 July 1967 memorandum of conversation (398), the 15 August 1967 memorandum for the record (418), the 9 October 1967 memorandum (463) from Rostow to Johnson, the 12 October 1967 memorandum (468) from Rostow to Johnson, the 13 October 1967 memorandum of conversation (473), and the 18 October 1967 memorandum of conversation (477).

42 *Ibid.*

4 Air Support: The 1968 Sale of F-4 Phantom Aircraft to Israel

1 See Warnke's 27 November 1968 letter (333) to Rabin. Also see the memoranda of conversation between American and Israeli officials dated 22 November 1968 (330) and 26 November 1968 (332). These documents are contained in *Foreign Relations of the United States, 1964–1968*, Volume XX, Arab–Israeli Dispute, 1967–68. All of the documents cited in this chapter are contained in this volume. (Each document's reference number in Volume XX appears in parentheses after its date of circulation.)

2 See Special Assistant to the President Walt Rostow's 24 December 1968 memorandum (360) to President Lyndon Johnson.

3 Israeli concern about the postwar air balance is reflected, explicitly or implicitly, in telegrams, memoranda, and letters dated 25 November 1967 (2), 27 November 1967 (3), 9 December 1967 (11), 11 December 1967 (15), 12 December 1967 (18), 15 December 1967 (20), 19 December 1967 (22), 7 January 1968 (38), 7 January 1968 (39), 8 January 1968 (40), and 8 January 1968 (41).

4 On the IAF's dwindling numbers of effective aircraft see American Ambassador to Israel Walworth Barbour's telegram to the DoS dated 27 November 1967 (3) and the memorandum of conversation between American and Israeli officials dated 8 January 1968 (40).

5 The numbers of aircraft requested by the Eshkol government are highlighted in Rostow's 7 January 1968 telegram (38) to Johnson. Also see the memo-

randa of conversations between American and Israeli officials dated 7 January 1968 (39) and 8 January 1968 (40).

6 These requests appear in telegrams, memoranda, and letters dated 25 November 1967 (2), 27 November 1967 (3), 12 December 1967 (18), 15 December 1967 (20), 19 December 1967 (22), 7 January 1968 (38), 7 January 1968 (39), 8 January 1968 (40), and 8 January 1968 (41).

7 See the memoranda and telegrams dated 11 December 1967 (15), 13 December 1967 (19), 15 December 1967 (20), 5 January 1968 (33), 6 January 1968 (36), 7 January 1968 (38), 8 January 1968 (40), and 8 January 1968 (41). See also the memoranda dated 3 May 1968 (164) and 21 May 1968 (179).

8 See the documents dated 7 January 1968 (38), 8 January 1968 (40), 3 May 1968 (164), and 21 May 1968 (179).

9 Johnson administration officials did confess, however, that their general assessment of the air balance would be invalidated if developments turned out to accord with Israeli rather than American expectations.

10 See the memoranda dated 11 December 1967 (15), 13 December 1967 (19), 15 December 1967 (20), 5 January 1968 (33), 6 January 1968 (36), and 7 January 1968 (38).

11 His opinion appears in his 27 November 1967 telegram (3) to the DoS.

12 The topics of conversation in these meetings are summarized in memoranda dated 7 January 1968 (39), 8 January 1968 (40), and 8 January 1968 (41).

13 See the memorandum dated 8 January 1968 (41).

14 See Rostow's 6 February 1968 memorandum (70) to Johnson.

15 See Rostow's 27 September 1968 memorandum (264) to Johnson. Clifford had replaced McNamara at the DoD in March.

16 See the memoranda dated 18 January 1968 (51) and 19 January 1968 (53).

17 See Rostow's 29 February 1968 memorandum (94) to Johnson.

18 See the 26 April 1968 memorandum of conversation (152) between American and Israeli officials.

19 Eshkol's letter to Johnson is contained in Rabin's 30 April 1968 letter (157) to Rusk. In a meeting with Rostow on 1 May 1968, Rabin reiterated the Israeli stand on the Phantom. See the 1 May 1968 memorandum of conversation (158) between American and Israeli officials.

20 See the 17 June 1968 memorandum of conversation (194) between American and Israeli officials, NSC member Harold Saunders's 6 August 1968 memorandum (231) to Johnson, the 9 September 1968 memorandum of conversation (247) between American and Israeli officials, Rusk's 1 October 1968 telegram (268) to the DoS, and the 22 October 1968 memorandum of conversation (284) between Johnson and Israeli Foreign Minister Abba Eban.

21 For American–Soviet diplomatic exchanges on an arms limitation agreement see the documents dated 17 January 1968 (48), 20 January 1968 (57), 22 January 1968 (58), 7 February 1968 (73), 27 February 1968 (93), 9 March 1968 (105), 17 May 1968 (176), and 4 September 1968 (245).

22 See the memoranda dated 26 April 1968 (152) and 21 May 1968 (179).

23 See Rostow's communications with Johnson dated 6 January 1968 (36) and 7 January 1968 (38), which reflected a general consensus within the administration.
24 Jarring had been appointed to facilitate talks on the basis of UNSCR 242.
25 See Saunders's 1 April 1968 memorandum (129) to Walt Rostow as well as Under Secretary of State for Political Affairs Eugene Rostow's 26 July 1968 memorandum (224) to Rusk. In June, prominent DoS officials, including Assistant Secretary of State for International Organization Affairs Joseph Sisco and Assistant Secretary of State for Near Eastern and South Asian Affairs Lucius Battle, wrote to Rusk urging the administration to provide the aircraft for this reason. See Sisco's 28 June 1968 memorandum (203) to Rusk.
26 See Rostow's 6 February 1968 memorandum (70) to Johnson.
27 See Rostow's 2 May 1968 memorandum (163) to Johnson and his 26 July 1968 memorandum (224) to Rusk.
28 Numerous documents in Volume XX grapple with the issue of arms for Jordan, including its connection to arms for Israel. These documents are dated as follows: 29 November 1967 (5), 6 December 1967 (9), 11 December 1967 (15), 13 December 1967 (19), 15 December 1967 (20), 26 December 1967 (24), 28 December 1967 (26), 30 December 1967 (29), 9 January 1968 (42), 17 January 1968 (49), 3 February 1968 (66), 6 February 1968 (71), 6 February 1968 (72), 10 February 1968 (76), 11 February 1968 (77), 14 February 1968 (82), 15 February 1968 (83), 20 February 1968 (89), 1 March 1968 (95), 12 March 1968 (107), 14 March 1968 (111), 26 March 1968 (125), 17 July 1968 (216), 17 September 1968 (253), 11 October 1968 (277), 12 October 1968 (278), 21 December 1968 (357), and 30 December 1968 (372).
29 The Johnson administration also agreed to unfreeze the delivery of the F-104 Starfighter, an aircraft that Jordan had purchased prior to the Six-Day War.
30 See Saunders's 14 October 1968 memorandum (279) to Rostow.
31 See the telegrams dated 28 April 1968 (155), 3 May 1968 (165), 2 July 1968 (205), 17 July 1968 (215), 11 September 1968 (250), and 1 October 1968 (268).
32 The Eshkol government had absolutely no intention of giving up the option to produce SSMs and nuclear weapons so long as the United States refused to provide Israel with an official security guarantee and so long as Egypt possessed weapons of mass destruction.
33 See Rusk's 24 October 1968 telegram (288) to the United States Embassy in Israel. For a very detailed review of this aspect of the Phantom negotiations see Avner Cohen, *Israel and the Bomb* (New York: Columbia University Press, 1998), pp. 303–321.
34 See the documents dated 25 October 1968 (290), 28 October 1968 (292), and 31 October 1968 (298).
35 See the memoranda dated 2 November 1968 (300) and 4 November 1968 (306).
36 Israel's stance is related in memoranda dated 4 November 1968 (306), 5

November 1968 (305), and 8 November 1968 (309).

37 See the memoranda dated 1 November 1968 (299) and 9 November 1968 (311).

38 For Johnson's pleas to the Eshkol government on the NPT see the documents dated 23 October 1968 (285) and 11 November 1968 (316).

39 See the 22 October 1968 memorandum of conversation (284).

40 For Johnson's knowledge of Israel's nuclear weapons see Cohen, *Israel and the Bomb*, pp. 306, 308.

41 See the documents dated 12 November 1968 (317), 22 November 1968 (330), 26 November 1968 (332), and 27 November 1968 (333).

42 *Ibid.*

5 National Interests or Domestic Politics?: The Rationale Behind American Arms Sales to Israel in the 1960s

1 The general tone of these arms negotiations comes through clearly in the memoranda of conversation between American and Soviet officials dated 17 February 1966 (272) and 16 March 1966 (280). See also Secretary of State Dean Rusk's and Secretary of Defense Robert McNamara's joint January–February 1966 memorandum (262) to President Lyndon Johnson and Deputy Special Assistant to the President for National Security Affairs Robert Komer's 8 February 1966 memorandum (267) to Johnson. See *Foreign Relations of the United States, 1964–1968*, Volume XVIII, Arab–Israeli Dispute, 1964–1967. All of the documents in this chapter are from this volume unless noted otherwise. (Each document's reference number in Volume XVIII appears in parentheses after its date of circulation.)

2 See the 16 March 1966 memorandum of conversation (280).

3 See the telegrams exchanged between Rusk and the United States Embassy in Egypt dated 29 February 1964 (22), 4 March 1964 (29), and 3 May 1964 (50).

4 These lines are taken from Komer's 8 February 1966 memorandum (267) to Johnson.

5 See American Ambassador to Israel Walworth Barbour's 11 March 1965 telegram (185) to the DoS.

6 For a thorough description and analysis of the concept of opacity see Avner Cohen, *Israel and the Bomb* (New York: Columbia University Press, 1998).

7 These lines are taken form Komer's 8 February 1966 memorandum (267) to Johnson.

8 The documents in Volume XVIII that highlight American–Israeli differences over the use of force are too numerous to cite individually.

9 See memorandum, Rusk to Johnson, 19 February 1965, "Israel, Tanks Vol. 2," Country File, National Security File, Box 145, LBJ Library.

10 See Komer's 8 February 1966 memorandum (267) to Johnson.

11 See Rusk's 30 July 1965 telegram (231) to Barbour.

12 *Ibid.*

13 These lines are taken from Rusk's and McNamara's January–February 1966 memorandum (262) to Johnson.

14 These lines come from Komer's 8 February 1966 memorandum (267) to Johnson.
15 On Johnson's warm ties to Eshkol see Aaron S. Klieman, *Israel & the World After 40 Years* (Washington, D.C.: Pergamon–Brassey's International Defense Publishers, 1990), pp. 74, 88. On Johnson's warm ties to Evron see Abba Eban, *Personal Witness: Israel Through My Eyes* (New York: G. P. Putnam's Sons, 1992), p. 479. On the latter relationship also see transcript, Harry McPherson Oral History Interview III, 16 January 1969 by T. H. Baker, Internet Copy, LBJ Library, pp. 16–17. McPherson inherited the job of liaison between the White House and the American Jewish community from Myer Feldman.
16 See National Security Council memorandum, 28 April 1964, "Israel, Tanks Vol. 1," Country File, National Security File, Box 145, LBJ Library.
17 See Rusk's and McNamara's January–February 1966 memorandum (262) to Johnson.
18 A survey of the domestic politics of the sale can be found in Steven L. Spiegel, *The Other Arab–Israeli Conflict: Making America's Middle East Policy, from Truman to Reagan* (Chicago: The University of Chicago Press, 1985), pp. 160–164.
19 For the administration's acknowledgment that firm public support for the sale existed in the United States see the 9 September 1968 notes (248) of Johnson's meeting with congressional leaders and Rusk's 9 October 1968 telegram (275) to American embassies in the Middle East. (The documents in this endnote and the following one are contained in *Foreign Relations of the United States, 1964–1968*, Volume XX, Arab–Israeli Dispute, 1967–1968.)
20 See Rusk's 9 October 1968 telegram (275) to American embassies in the region.
21 On Johnson's solicitation of the presidential candidates see Eban, *Personal Witness*, p. 474.

6 The 1967 Six-Day War: A Delayed "Green Light" for Preemption

1 The following sources trace American and Israeli conduct in the weeks before the outbreak of the war: Michael Brecher, *Decisions in Crisis: Israel, 1967 and 1973* (Berkeley: University of California Press, 1980); Abba Eban, *Personal Witness: Israel Through My Eyes* (New York: G.P. Putnam's Sons, 1992); Michael B. Oren, *Six Days of War: June 1967 and the Making of the Modern Middle East* (New York: Oxford University Press, 2002); and Steven L. Spiegel, *The Other Arab–Israeli Conflict: Making America's Middle East Policy, from Truman to Reagan* (Chicago: The University of Chicago Press, 1985), pp. 136–150.
2 See the 17 May 1967 telegram (8) to the United States Embassy in Israel. This document is contained in *Foreign Relations of the United States, 1964–1968*, Volume XIX, Arab–Israeli Crisis and War, 1967. All of the documents cited in this chapter are contained in this volume. (Each document's reference number in Volume XIX appears in parentheses after its date of circulation.)

3 See the 18 May 1967 telegram (13) to the Department of State (DoS).
4 Eban, *Personal Witness*, p. 408. In the same vein, Foreign Minister Abba Eban told fellow cabinet ministers that, "The question is not whether we must resist, but whether we must resist alone or with the support and understanding of others. . . . Otherwise, we may win a war and lose a victory." Quoted in Brecher, *Decisions in Crisis*, p. 120.
5 See the 21 May 1967 telegram (30) to the United States Embassy in Israel.
6 See the 25 and 26 May 1967 memoranda of conversations (64, 69, and 77) between American and Israeli officials, the 26 May 1967 memorandum (71) from Secretary of State Dean Rusk to President Lyndon Johnson, and the 27 May 1967 telegram (86) to the United States Embassy in Israel.
7 Eban, *Personal Witness*, p. 399. See also the 29 and 30 May 1967 telegrams (97 and 98) between the DoS and the United States Embassy in Israel.
8 Quoted in Brecher, *Decisions in Crisis*, p. 147. See also Eshkol's 30 May 1967 letter (102) to Johnson.
9 See the 1 June 1967 memorandum for the record (124).
10 Eban, *Personal Witness*, p. 405.
11 For these predictions see the 23 and 26 May 1967 Central Intelligence Agency (CIA) memoranda (44 and 76), the 30 May 1967 and 2 June 1967 memoranda of conversations (99 and 130), and the 3 June 1967 memorandum (142) from National Security Council Staff Member Robert Ginsburg to the President's Special Assistant Walt Rostow.
12 On the potential for damage to American national interests in the Middle East see the 31 May 1967 memo (114) from National Security Council Staff Member Harold Saunders to Rostow, the 31 May 1967 report of the Working Group on Economic Vulnerability (115), the 1 June 1967 memorandum (126) from the Board of National Estimates to CIA Director Richard Helms, and the 3 June 1967 CIA memorandum (143).
13 For the connection between the Middle East and Vietnam conflicts see Judith A. Klinghoffer, *Vietnam, the Jews, and the Middle East: Unintended Consequences* (New York: Palgrave Macmillan, 1999).
14 For accounts of American conduct during and after the Six-Day War see David Schoenbaum, *The United States and the State of Israel* (New York: Oxford University Press, 1993), pp. 154–169 and Spiegel, *The Other Arab–Israeli Conflict*, pp. 150–164.

7 The 1969–1970 War of Attrition: Restricting Israel's Military Options

1 Egypt's attitude, noted Israeli Foreign Minister Abba Eban, could be summed up as "that which was taken by force must be recaptured by force." See Abba Eban, *Personal Witness: Israel Through My Eyes* (New York: G.P. Putnam's Sons, 1992), p. 484.
2 On the late October 1968 round of fighting see Jonathan Shimshoni, *Israel and Conventional Deterrence: Border Warfare from 1953 to 1970* (Ithaca, NY: Cornell University Press, 1988), pp. 138–141.
3 For an account of an Israel Air Force (IAF) photo reconnaissance sortie over

the Aswan High Dam see Amos Amir, *Fire in the Sky: Flying in Defence of Israel* (South Yorkshire: Pen and Sword Books, 2005), pp. 46–56.

4 For American and Israeli conduct during the War of Attrition see Yaacov Bar-Siman-Tov, *The Israeli–Egyptian War of Attrition, 1969–1970: A Case Study of Limited Local War* (New York: Columbia University Press, 1980); David Pollock, *The Politics of Pressure: American Arms and Israeli Policy Since the Six Day War* (Westport, CT: Greenwood Press, 1982), pp. 57–102; Shimshoni, *Israel and Conventional Deterrence*, pp. 123–211; and Avi Shlaim and Raymond Tanter, "Decision, Process, Choice, and Consequences: Israel's Deep-Penetration Bombing in Egypt, 1970," *World Politics*, Vol. 30, No. 4 (July 1978), pp. 483–516.

5 See Bar-Siman-Tov, *The Israeli–Egyptian War of Attrition*, pp. 134–135 for a list of deep-penetration air raids during the months of January–April.

6 For the numbers of Soviet troops deployed to Egypt during the War of Attrition see Jon D. Glassman, *Arms for the Arabs: The Soviet Union and War in the Middle East* (Baltimore, MD: The Johns Hopkins University Press, 1975), p. 75 and Efraim Karsh, *The Cautious Bear: Soviet Military Engagement in Middle Eastern Wars in the Post-1967 Era* (Boulder, CO: Westview Press, 1987), pp. 16, 64–65, 77.

7 Eban, *Personal Witness*, p. 484.

8 For American opposition to the Israeli air raids see Bar-Siman-Tov, *The Israeli–Egyptian War of Attrition*, p. 157; Pollock, *The Politics of Pressure*, pp. 66–72; and Shlaim and Tanter, "Decision, Process, Choice, and Consequences," pp. 501–506.

9 Quoted in Pollock, *The Politics of Pressure*, p. 68.

10 On American efforts to get Israel to accept a cease-fire see Bar-Siman-Tov, *The Israeli–Egyptian War of Attrition*, pp. 181–185 and Pollock, *The Politics of Pressure*, pp. 72–86.

11 Eban, *Personal Witness*, p. 489.

12 *Ibid.*, p. 490.

13 The fact that Israel played a key role in deterring a full-scale Syrian intervention on behalf of Palestinian forces seeking the overthrow of Jordan's pro-Western regime during the Jordanian civil war of September 1970 also contributed to American willingness to shower the Jewish state with arms.

14 Bar-Siman-Tov, *The Israeli–Egyptian War of Attrition*, pp. 130–131.

8 The 1973 Yom Kippur War: Limiting Israel's Military Victory

1 The most authoritative account of why Israel did not predict the outbreak of war, in spite of all of the information in its hands, can be found in Uri Bar-Joseph, *The Watchman Fell Asleep: The Surprise of Yom Kippur and Its Sources* (Albany, NY: State University of New York Press, 2005).

2 For accounts of American and Israeli conduct in the war see Michael Brecher, *Decisions in Crisis: Israel, 1967 and 1973* (Berkeley: University of California Press, 1980), pp. 51–76, 171–229, 286–324; David Pollock, *The Politics of Pressure: American Arms and Israeli Policy Since the Six Day War*

(Westport, CT: Greenwood Press, 1982), pp. 157–216; and Steven L. Spiegel, *The Other Arab–Israeli Conflict: Making America's Middle East Policy, from Truman to Reagan* (Chicago: The University of Chicago Press, 1985), pp. 219–314.

3 On the American role in the Meir government's decision to forgo a preemptive air attack see Brecher, *Decisions in Crisis*, pp. 177–178, 187–188, 197–201 and Pollock, *The Politics of Pressure*, pp. 170–172.

4 See the 7 October 1973 memorandum of conversation between Secretary of State Henry Kissinger and Israeli Ambassador to the United States Simcha Dinitz in "Records of Henry Kissinger, 1973–1977," Record Group 59, Department of State (DoS) Records, Box 25 (Category C 1973 Arab–Israeli War), National Archives.

5 See the 6 October 1973 telegram from Kissinger to President Richard Nixon in "Middle East War Memos and Miscellaneous (1 October–17 October 1973)," National Security Council (NSC) File, Nixon Presidential Materials Project, Box 664, National Archives.

6 On the initial American response to the Jewish state's arms requests see Brecher, *Decisions in Crisis*, pp. 209, 213–216 and Pollock, *The Politics of Pressure*, pp. 172–174. For the transfer of American arms in Israeli planes see the 7 October 1973 memorandum of conversation between Kissinger and Dinitz in "Records of Henry Kissinger, 1973–1977," Record Group 59, DoS Records, Box 25 (Category C 1973 Arab–Israeli War), National Archives and the 9 October 1973 memorandum of conversation between them in "Subject–Numeric Files 70–73, Policy Israel–United States," Record Group 59, DoS Records, National Archives.

7 See the 21 October 1973 telegram from Kissinger to General Brent Scowcroft, his deputy, in "Henry A. Kissinger Trip–Moscow, Tel Aviv, London (20 October–23 October 1973)," Henry Kissinger Office Files, Nixon Presidential Materials Project, Box 39, National Archives and the 22 October 1973 memorandum of conversation between Kissinger and Prime Minister Golda Meir in "Subject–Numeric Files 70–73, Policy 7 United States–Kissinger," Record Group 59, DoS Records, National Archives.

8 On American objections to the destruction of the Third Army see Brecher, *Decisions in Crisis*, pp. 226–227, 296–298, 302–304 and Pollock, *The Politics of Pressure*, pp. 177–178. To get a flavor of the discussions that took place between the United States and Israel on this issue see the 1 November 1973 memoranda of conversations among Kissinger, Nixon, and Meir in "Records of Henry Kissinger, 1973–1977," Record Group 59, DoS Records, Box 2 (NODIS Action Memos, 1973–1976), National Archives.

9 A high-ranking Soviet official, incidentally, later revealed that the Soviet Union had not given serious thought to direct military intervention on behalf of Egypt. Victor Israelyan, *Inside the Kremlin During the Yom Kippur War* (University Park, PA: Pennsylvania State University Press, 1995).

10 Abba Eban, *Personal Witness: Israel Through My Eyes* (New York: G.P. Putnam's Sons, 1992), p. 538.

11 For American pressure during these postwar talks see Pollock, *The Politics of Pressure*, pp. 179–196.
12 Quoted in Spiegel, *The Other Arab–Israeli Conflict*, pp. 248–249.
13 See the 23 October 1973 briefing by Kissinger to his staff in "Transcripts of Secretary of State Henry A. Kissinger Staff Meetings, 1973–1977," Record Group 59, DoS Records, Box 1, National Archives.
14 See the 17 October 1973 memorandum of conversation between Nixon and his advisors in "Washington Special Action Group Principles: Middle East War (17 October)," NSC Institutional File, Box H–92 (Folder 6), National Archives.
15 Though calling for a preemptive attack in order to improve its prospects at the outset of the war, the IDF had also assured the government that Israel would eventually emerge victorious no matter which side fired the opening salvo. The IDF's confidence that it would win the war under any set of circumstances served to buttress the government's reasoning that Israel was correct to surrender to American pressure on the issue of a preemptive strike.
16 Moshe Dayan, *Moshe Dayan: Story of My Life* (New York: Warner Books, 1976), p. 556.
17 Quoted in Brecher, *Decisions in Crisis*, p. 181.

9 Peacetime Arms Transfers: The Nixon, Carter, and Reagan Administrations

1 For accounts of the American–Israeli relationship during the Nixon administration between the 1969–1970 War of Attrition and the 1973 Yom Kippur War see David Pollock, *The Politics of Pressure: American Arms and Israeli Policy Since the Six Day War* (Westport, CT: Greenwood Press, 1982), pp. 78–86, 103–155 and Steven L. Spiegel, *The Other Arab–Israeli Conflict: Making America's Middle East Policy, from Truman to Reagan* (Chicago: The University of Chicago Press, 1985), pp. 166–218.
2 A review of the events surrounding the civil war can be found in Paul K. Huth, *Extended Deterrence and the Prevention of War* (New Haven: Yale University Press, 1988), pp. 86–97.
3 Pollock, *The Politics of Pressure*, p. 84.
4 Early in the Nixon administration's tenure in office, some government officials made one last attempt to convince the White House to use arms transfers as a lever to dissuade Israel from moving forward with its nuclear weapons and ballistic missile programs. President Richard Nixon and National Security Advisor (and, later, Secretary of State) Henry Kissinger, however, rejected their advice, instead opting to solidify the tacit agreement between the United States and the Jewish state reached under the preceding Johnson administration whereby the latter would keep its nuclear weapons and ballistic missile programs out of the public eye in exchange for American arms. See Avner Cohen and William Burr, "Israel Crosses the Threshold," *Bulletin of the Atomic Scientists*, Vol. 62, No. 3 (May/June 2006), pp. 22–30, which is based on recently declassified American government documents.

5 Pollock, *The Politics of Pressure*, p. 126.
6 *Ibid.*, pp. 128–130.
7 For accounts of the American–Israeli relationship during the Carter administration see Pollock, *The Politics of Pressure*, pp. 217–273 and Spiegel, *The Other Arab–Israeli Conflict*, pp. 315–380.
8 Pollock, *The Politics of Pressure*, p. 236. The proposed Geneva Peace Conference, however, never took place. Egyptian President Anwar el-Sadat's dramatic visit to the Jewish state spawned separate Israeli–Egyptian peace negotiations that rendered an international peace conference moot.
9 *Ibid.*, pp. 234–236.
10 *Ibid.*, pp. 242–244.
11 For accounts of this arms sale see Pollock, *The Politics of Pressure*, pp. 237–241 and Spiegel, *The Other Arab–Israeli Conflict*, pp. 346–349.

12 Secretary of State Cyrus Vance's 14 February 1978 remarks may be found on the Israel Ministry of Foreign Affairs web site <www.mfa.gov.il>.
13 President Jimmy Carter's 19 February 1978 disavowal may also be found on the Ministry of Foreign Affairs web site <www.mfa.gov.il>.
14 Zbigniew Brzezinski, *Power and Principle: Memoirs of the National Security Adviser* (New York: Farrar, Straus, & Giroux, 1983), p. 248.
15 Prime Minister Menachem Begin's 15 February 1978 comments can be found on the Israel Ministry of Foreign Affairs web site <www.mfa.gov.il>.
16 Pollock, *The Politics of Pressure*, pp. 241–242, 247–249.
17 For accounts of the American–Israeli relationship during the first Reagan administration see Pollock, *The Politics of Pressure*, pp. 275–292 and Spiegel, *The Other Arab–Israeli Conflict*, pp. 395–429. The 1982 Lebanon War will not be examined here, as the purpose of the present discussion is to probe the security-for-autonomy bargain in the absence of full-scale war.
18 Pollock, *The Politics of Pressure*, pp. 287–289.
19 Quoted in A. F. K. Organski, *The $36 Billion Bargain: Strategy and Politics in U.S. Assistance to Israel* (New York: Columbia University Press, 1990), p. 188.
20 Begin's 20 December 1981 remarks can be found on the Israel Ministry of Foreign Affairs web site <www.mfa.gov.il>.
21 Pollock, *The Politics of Pressure*, pp. 278–279, 282–284.
22 The 29 October 1981 cabinet statement can be found on the Israel Ministry of Foreign Affairs web site <www.mfa.gov.il>.
23 Pollock, *The Politics of Pressure*, pp. 279–282.

Conclusion: The Costs of an Alliance and the Benefits of a Patron–Client Relationship

1 Aaron S. Klieman, *Israel and the World after 40 Years* (Washington, D.C.: Pergamon–Brassey's International Defense Publishers, Inc., 1990), p. 199 and Michael Mandelbaum, *The Fate of Nations: The Search for National Security in the Nineteenth and Twentieth Centuries* (New York: Cambridge University Press, 1988), p. 308.
2 Israel, to be sure, showed great interest in an American security guarantee

during the early decades of its existence; but, in seeking such a guarantee, it did not have a constraining alliance in mind. Instead, it sought an American commitment to come to the Jewish state's aid in an hour of need without any reciprocal Israeli obligation to the United States.

3 For a general review of the concept of entrapment in alliances see Glenn H. Snyder, "The Security Dilemma in Alliance Politics," *World Politics*, Vol. 36, No. 4 (July 1984), pp. 461–495.

4 For a general history of Israel's intelligence services, including links to their American counterparts, see Dan Raviv and Yossi Melman, *Every Spy a Prince: The Complete History of Israel's Intelligence Community* (Boston: Houghton Mifflin, 1990).

5 An overview of American–Israeli military cooperation through the mid-1980s may be found in Wolf Blitzer, *Between Washington and Jerusalem: A Reporter's Notebook* (New York: Oxford University Press, 1985).

6 For a general review of the concept of abandonment in alliances see Snyder, "The Security Dilemma." See also Klieman, *Israel and the World*, p. 199 and Mandelbaum, *The Fate of Nations*, p. 308.

7 Mandelbaum, *The Fate of Nations*, p. 308 and Gerald L. Sorokin, "Patrons, Clients, and Allies in the Arab–Israeli Conflict," *The Journal of Strategic Studies*, Vol. 20, No. 1 (March 1997), pp. 46–71.

Select Bibliography

Articles

Cohen, Avner and William Burr, "Israel Crosses the Threshold," *Bulletin of the Atomic Scientists*, Vol. 62, No. 3 (May/June 2006), pp. 22–30.

Morrow, James D., "Alliances and Asymmetry: An Alternative to the Capability Aggregation Model of Alliances," *American Journal of Political Science*, Vol. 35, No. 4 (November 1991).

Rodman, David, "Patron–Client Dynamics: Mapping the American–Israeli Relationship," *Israel Affairs*, Vol. 4, No. 2 (Winter 1997).

Shlaim, Avi and Raymond Tanter, "Decision, Process, Choice, and Consequences: Israel's Deep-Penetration Bombing in Egypt, 1970," *World Politics*, Vol. 30, No. 4. (July 1978).

Slonim, Shlomo, "The 1948 American Embargo on Arms to Palestine," *Political Science Quarterly*, Vol. 94, No. 3 (Fall 1979).

Snyder, Glenn H., "The Security Dilemma in Alliance Politics," *World Politics*, Vol. 36, No. 4 (July 1984).

Sorokin, Gerald L., "Patrons, Clients, and Allies in the Arab–Israeli Conflict," *The Journal of Strategic Studies*, Vol. 20, No. 1 (March 1997).

Books

Almog, Orna, *Britain, Israel, and the United States, 1955–1958: Beyond Suez* (London: Frank Cass, 2003).

Alteras, Isaac, *Eisenhower and Israel: U.S.–Israeli Relations, 1953–1956* (Gainesville, FL: University of Florida Press, 1993).

Amir, Amos, *Fire in the Sky: Flying in Defence of Israel* (South Yorkshire: Pen and Sword Books, 2005).

Bar-Joseph, Uri, *The Watchman Fell Asleep: The Surprise of Yom Kippur and Its Sources* (Albany, NY: State University of New York Press, 2005).

Bar-On, Mordechai, *The Gates of Gaza: Israel's Road to Suez and Back, 1955–1957* (New York: St. Martin's Press, 1994).

Bar-Siman-Tov, Yaacov, *The Israeli–Egyptian War of Attrition, 1969–1970: A Case Study of Limited Local War* (New York: Columbia University Press, 1980).

Bass, Warren, *Support Any Friend: Kennedy's Middle East and the Making of the U.S.–Israel Alliance* (New York: Oxford University Press, 2003).

Ben-Zvi, Abraham, *Decade of Transition: Eisenhower, Kennedy, and the Origins of the American–Israeli Alliance* (New York: Columbia University Press, 1998).

Ben-Zvi, Abraham, *John F. Kennedy and the Politics of Arms Sales to Israel* (London: Frank Cass, 2002).
Ben-Zvi, Abraham, *Lyndon B. Johnson and the Politics of Arms Sales to Israel: In the Shadow of the Hawk* (London: Frank Cass, 2004).
Ben-Zvi, Abraham, *The United States and Israel* (New York: Columbia University Press, 1993).
Bercuson, David J., *The Secret Army* (New York: Stein and Day, 1984).
Bialer, Uri, *Between East and West: Israel's Foreign Policy Orientation* (Cambridge: Cambridge University Press, 1990).
Blitzer, Wolf, *Between Washington and Jerusalem: A Reporter's Notebook* (New York: Oxford University Press, 1985).
Brecher, Michael, *Decisions in Crisis: Israel, 1967 and 1973* (Berkeley: University of California Press, 1980).
Brecher, Michael, *Decisions in Israel's Foreign Policy* (New Haven: Yale University Press, 1975).
Brzezinski, Zbigniew, *Power and Principle: Memoirs of the National Security Advisor* (New York: Farrar, Strauss, and Giroux, 1983).
Cohen, Avner, *Israel and the Bomb* (New York: Columbia University Press, 1998).
Cohen, Michael J., *Fighting World War III from the Middle East: Allied Contingency Plans, 1945–1954* (London: Frank Cass, 1997).
Cohen, Michael J., *Truman and Israel* (Berkeley: University of California Press, 1990).
Crosbie, Sylvia K., *A Tacit Alliance: France and Israel from Suez to the Six-Day War* (Princeton, NJ: Princeton University Press, 1974).
Dayan, Moshe, *Moshe Dayan: Story of My Life* (New York: Warner Books, 1976).
Eban, Abba, *Israel Through My Eyes* (New York: G. P. Putnam's Sons, 1992).
Eban, Abba, *The New Diplomacy: International Affairs in the Modern Age* (New York: Random House, 1983).
Feldman, Lily Gardner, *The Special Relationship Between West Germany and Israel* (Boston: Allen & Unwin, 1984).
Gazit, Mordechai, *President Kennedy's Policy Toward the Arab States and Israel: Analysis and Documents* (Tel Aviv: Shiloah Center for Middle Eastern and African Studies, 1983).
Glassman, Jon D., *Arms for the Arabs: The Soviet Union and War in the Middle East* (Baltimore, MD: The Johns Hopkins University Press, 1975).
Golani, Motti, *Israel in Search of a War: The Sinai Campaign, 1955–1956* (Brighton & Portland: Sussex Academic Press, 1998).
Hahn, Peter L., *Caught in the Middle: U.S. Policy Toward the Arab–Israeli Conflict, 1945–1961* (Chapel Hill, NC: The University of North Carolina Press, 2004).
Heckelman, Joseph A., *American Volunteers and Israel's War of Independence* (New York: KTAV Publishing House, 1974).
Holsti, Ole, P. Terrance Hopmann, and John D. Sullivan, *Unity and Disintegration in International Alliances* (New York: John Wiley & Sons, 1973).
Huth, Paul K., *Extended Deterrence and the Prevention of War* (New Haven: Yale University Press, 1988).
Ilan, Amitzur, *The Origin of the Arab–Israeli Arms Race: Arms, Embargo, Military*

Power and Decision in the 1948 Palestine War (New York: New York University Press, 1996).

Israelyan, Victor, *Inside the Kremlin During the Yom Kippur War* (University Park, PA: Pennsylvania State University Press, 1995).

Karsh, Efraim, *The Cautious Bear: Soviet Military Engagement in Middle Eastern Wars in the Post-1967 Era* (Boulder, CO: Westview Press, 1987).

Kegley, Jr., Charles W., and Gregory A. Raymond, *When Trust Breaks Down: Alliance Norms and World Politics* (Columbia, SC: University of South Carolina Press, 1990).

Klieman, Aaron S., *Israel & the World After 40 Years* (Washington, D.C.: Pergamon–Brassey's International Defense Publishers, Inc., 1990).

Klinghoffer, Judith A., *Vietnam, the Jews, and the Middle East: Unintended Consequences* (New York: Palgrave Macmillan, 1999).

Levey, Zach, *Israel and the Western Powers, 1952–1960* (Chapel Hill, NC: The University of North Carolina Press, 1997).

Mandelbaum, Michael, *The Fate of Nations: The Search for National Security in the Nineteenth and Twentieth Centuries* (New York: Cambridge University Press, 1988).

Oren, Michael B., *Six Days of War: June 1967 and the Making of the Modern Middle East* (New York: Oxford University Press, 2002).

Organski, A. F. K., *The $36 Billion Bargain: Strategy and Politics in U.S. Assistance to Israel* (New York: Columbia University Press, 1990).

Pollock, David, *The Politics of Pressure: American Arms and Israeli Policy Since the Six Day War* (Westport, CT: Greenwood Press, 1982).

Raviv, Dan and Yossi Melman, *Every Spy a Prince: The Complete History of Israel's Intelligence Community* (Boston: Houghton Mifflin, 1990).

Reiser, Stewart, *The Israeli Arms Industry: Foreign Policy, Arms Transfers, and Military Doctrine of a Small State* (New York: Holmes & Meier, 1989).

Schoenbaum, David, *The United States and the State of Israel* (New York: Oxford University Press, 1993).

Shalom, Zaki, *Israel's Nuclear Option: Behind the Scenes Diplomacy Between Dimona and Washington* (Brighton & Portland: Sussex Academic Press, 2005).

Shalom, Zaki, *The Superpowers, Israel and the Future of Jordan, 1960–1963* (Brighton & Portland: Sussex Academic Press, 1999).

Shimshoni, Jonathan, *Israel and Conventional Deterrence: Border Warfare from 1953 to 1970* (Ithaca, NY: Cornell University Press, 1988).

Slater, Leonard, *The Pledge* (New York: Simon & Schuster, 1970).

Spiegel, Steven L., *The Other Arab–Israeli Conflict: Making America's Middle East Policy, from Truman to Reagan* (Chicago: The University of Chicago Press, 1985).

Walt, Stephen M., *The Origins of Alliances* (Ithaca, NY: Cornell University Press, 1987).

Weiss, Jeffrey and Craig Weiss, *I Am My Brother's Keeper: American Volunteers in Israel's War for Independence* (Atglen, PA: Schiffer Military History, 1998).

Yaniv, Avner, *Deterrence Without the Bomb: The Politics of Israeli Strategy* (Lexington, MA: Lexington Books, 1987).

Documents

The Avalon Project at the Yale Law School <www.yale.edu/lawweb/avalon>.
Foreign Relations of the United States, 1964–1968, Volume XVIII, Arab–Israeli Dispute, 1964–67 (Washington, D.C.: United States Government Printing Office, 2000).
Foreign Relations of the United States, 1964–1968, Volume XIX, Arab–Israeli Crisis and War, 1967 (Washington, D.C.: United States Government Printing Office, 2004).
Foreign Relations of the United States, 1964–1968, Volume XX, Arab–Israeli Dispute, 1967–68 (Washington, D.C.: United States Government Printing Office, 2001).
Israel Ministry of Foreign Affairs <www.mfa.gov.il>, historical documents (various years).
Lyndon Baines Johnson Presidential Library (various boxes and files).
United States National Archives (various boxes and files).

Index